1

Learning to Analyze a Power Supply

This was the "IRA760" which I built for Siemens AG in Germany (1995-96)! (Forward converter because of its choke)

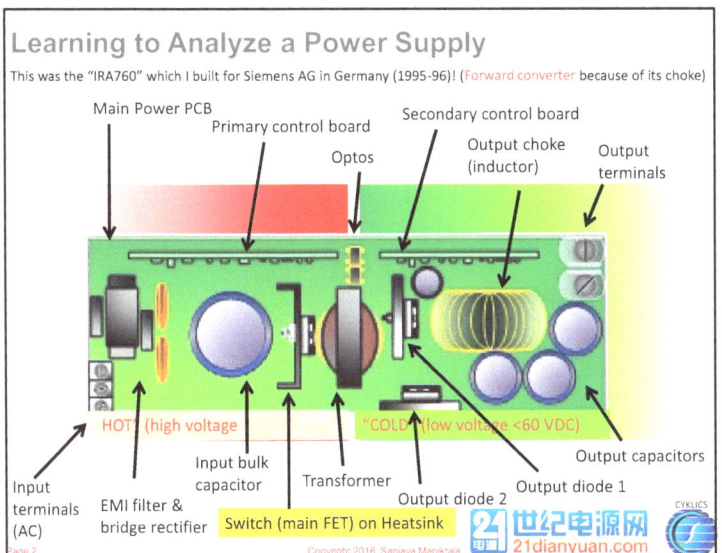

2

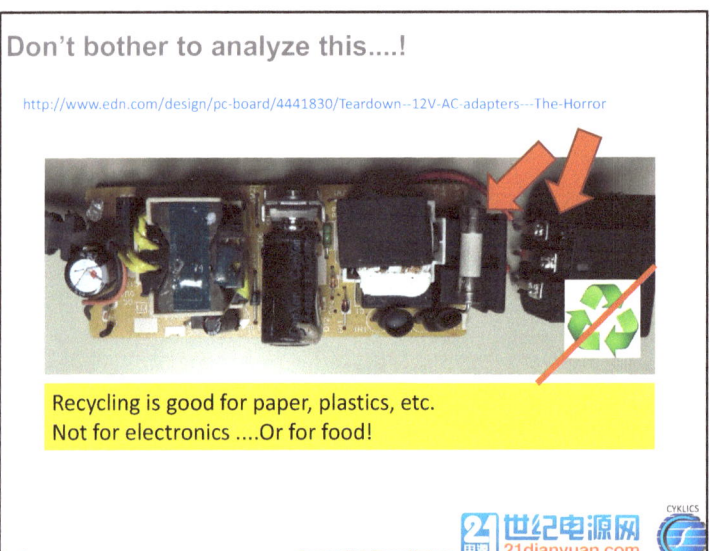

Contents

5

6

3

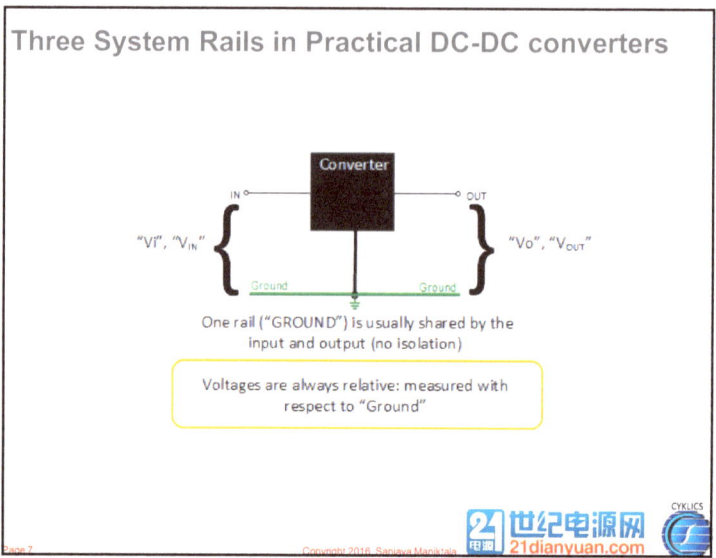

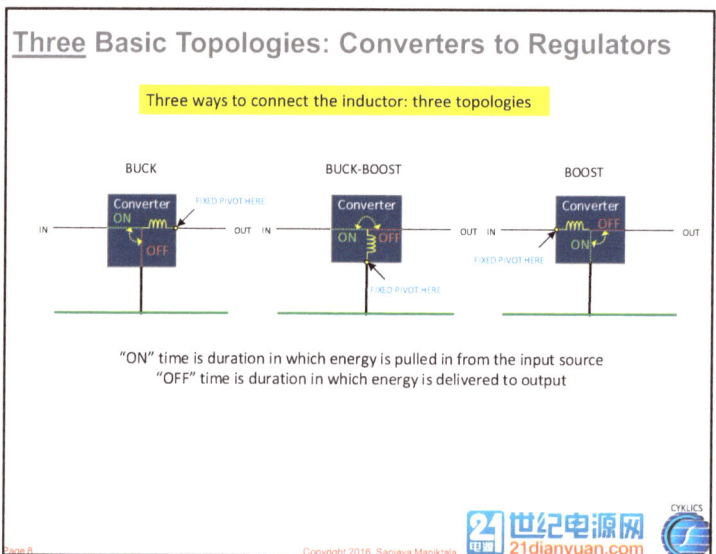

Three Basic Topologies: Converters to Regulators

❑One end of the inductor is always connected to a *fixed voltage rail*…

❑The other end is switched between the other two rails repetitively… at a certain "switching frequency"….

❑Switching the FET reduces dissipation in the FET because either voltage across it is "zero" (ON-time), or current through it is "zero" (OFF-time)…theoretically V × I is "zero" all the time…

❑This switching across an inductor produces an induced voltage (Faraday's law) which can be *made more suitable* than the one we started off with (input source)…

❑We can control/regulate the output voltage by "pulse width modulation (PWM)" (explained later)

9

The Buck Topology

unstoppable.

Switch (FET) goes ON and OFF repetitively

"Switching node"   "Fixed node"

IN○ ○ ON OUT
OFF

Important: Inductor current cannot jump suddenly! It "freewheels"

Inductor current "freewheels" through diode when Switch turns OFF

COMMON GROUND   COMMON GROUND

Fixed node is connected to Output rail

Step-down (Buck)

Inductor current direction and magnitude must be maintained at each switch transition

10

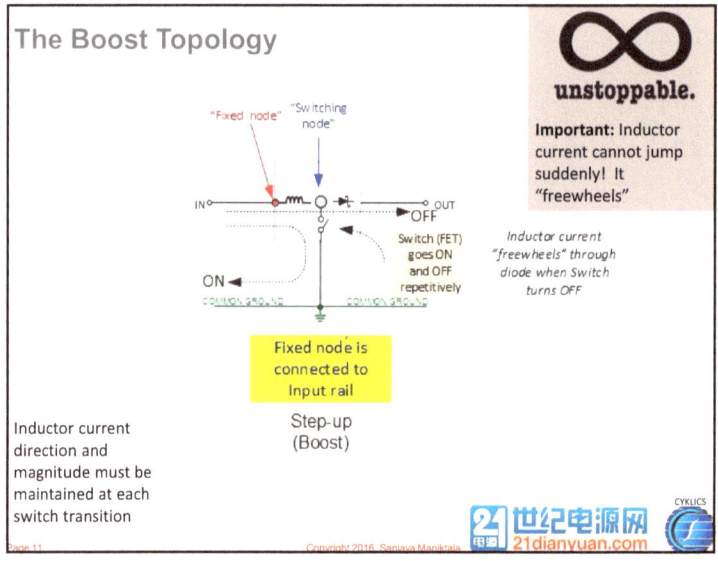

11

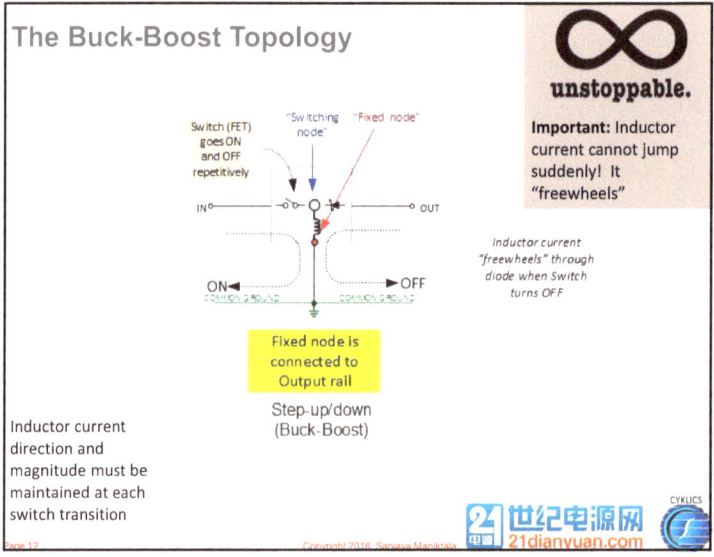

12

DC/DC converters/regulators

- **Three** basic, fundamental topologies (i.e. using **one** inductor). They behave as follows:

- **Buck** (Step-down): 12V to 5V, -12V to -5V (sign of output same as sign of input: "non-inverting")
- **Boos**t (Step-up): 5V to 12V, -5V to -12V (sign of output same as sign of input: "non-inverting"))
- **Buck-Boost** (Step-up or Step-down): 12V to -5V, -12V to 5V, 5V to -12V, -5V to 12V, 5V to -5V, 12V to -12V, -5V to 5V, -12V to 12V (sign of output is opposite to sign of input: this is an "inverting" topology)

- The words "step-up (boost)" or "step-down (buck)" refer only to the *magnitudes* of the input and output voltages.

Copyright 2016, Sanjaya Maniktala

13

Ideal Duty Equations: Connect Vo to Vin

CCM assumed	Buck	Boost	Buck-Boost
Duty cycle "D"	$\approx \dfrac{V_O}{V_{IN}}$	$\approx \dfrac{V_O - V_{IN}}{V_O}$	$\approx \dfrac{V_O}{V_{IN} + V_O}$

Keep in mind that these "DC transfer function" equations are for "continuous conduction mode (CCM)"....in which *the inductor current never returns to zero in a switching cycle*. Also the sign of the output and input is being ignored here.

For a given D, the output voltage is a fixed fraction/ratio of the input voltage... if input falls, all voltages in the converter fall, if input rises, all the voltages rise ...it is quite natural. Hence the need for closed loop correction, to increase D if input falls.

Note: these do not involve load current ....so does that mean we can get any load current for a given D, Vin and Vo? Partly true. Discussed later!

Copyright 2016, Sanjaya Maniktala

14

15

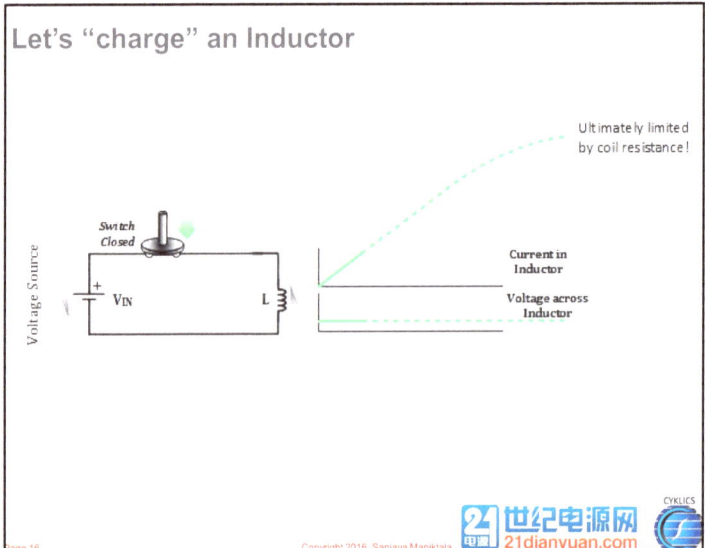

16

Question: What happens

- **If the switch is opened after a short while?**

- What happens to the inductor current?

- How can it just "go away somewhere"?

- Have you heard of the law of *conservation of energy*?

- Doesn't the inductor have stored energy prior to opening the switch?

- Where does that energy "disappear"?

17

"Discharging" the Inductor

A manifestation of the Law of Conservation of Energy: *Energy can only be be converted from one form to another (or eventually lost as heat)*

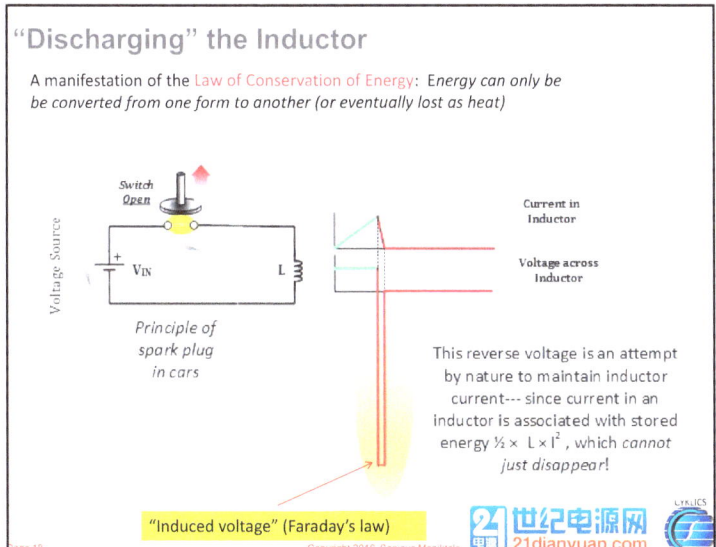

18

L-C analogy (duality): It's easier to visualize

- When we discharge a capacitor, we get a very large spike of current **through it**

(electric stored energy is dumped as heat in the undocumented/parasitic resistances)

- When we "discharge" an inductor, we get a very large spike of voltage **across it**

(magnetic stored energy is dumped as heat in the undocumented/parasitic resistances)

> Keep in mind that in a "perfect" inductor or capacitor, we can never dissipate any energy...these are **reactive** (energy storing) components. To dissipate energy as heat, there must be a **resistance present somewhere!**

19

The "Inductor Equation" applies during both ON and OFF times

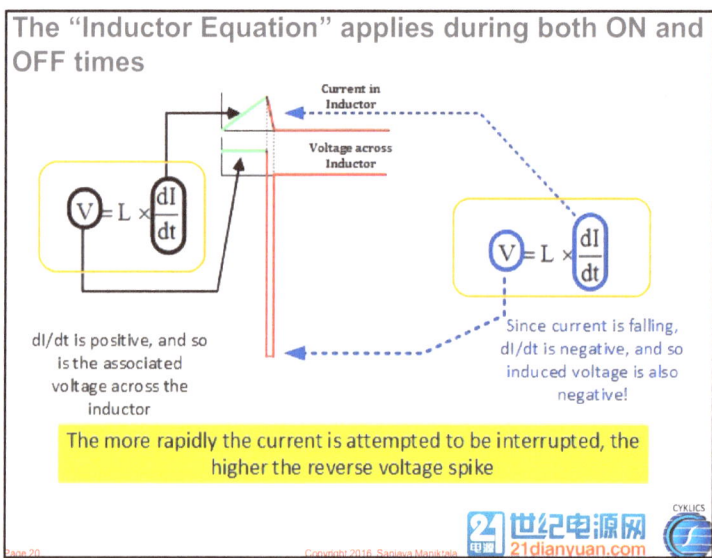

$$V = L \times \frac{dI}{dt}$$

$$V = L \times \frac{dI}{dt}$$

Current in Inductor

Voltage across Inductor

dI/dt is positive, and so is the associated voltage across the inductor

Since current is falling, dI/dt is negative, and so induced voltage is also negative!

The more rapidly the current is attempted to be interrupted, the higher the reverse voltage spike

20

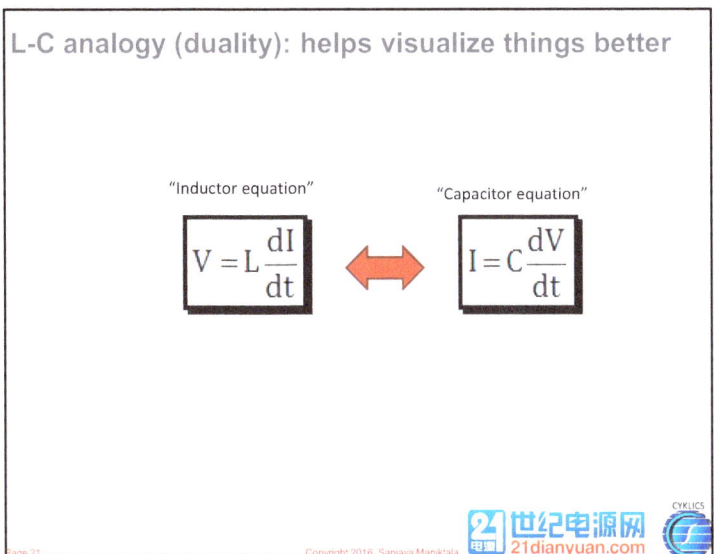

21

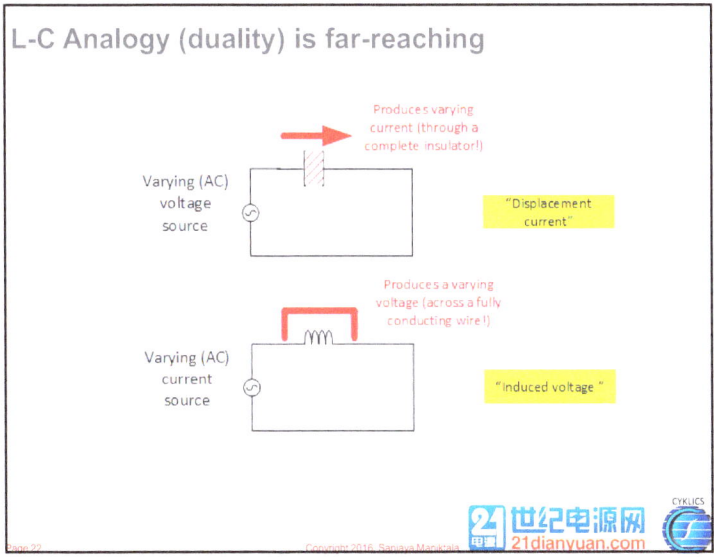

22

Duality Principles

Capacitors can help us understand inductors, and vice versa…!

https://en.wikipedia.org/wiki/Duality_%28electricity_and_magnetism%29

- The electric field (**E**) is the dual of the magnetizing field (**H**)
- The electric displacement field (**D**) is the dual of the magnetic flux density (**B**)
- Faraday's law of induction is the dual of Ampere's circuital law
- Gauss's law for electric field is the dual of Gauss's law for magnetism
- The electric potential is the dual of the magnetic potential
- Permittivity is the dual of permeability
- Electrostriction is the dual of magnetostriction
- Piezoelectricity is the dual of piezomagnetism
- Ferroelectricity is the dual of ferromagnetism
- An electrostatic motor is the dual of a magnetic motor
- Electrets are the dual of permanent magnets
- The Faraday effect is the dual of the Kerr effect
- The Aharonov–Casher effect is the dual to the Aharonov–Bohm effect
- The hypothetical magnetic monopole is the dual of electric charge

Read more if interested! Skip for now

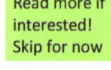

Page 23 Copyright 2016, Sanjaya Maniktala

23

Duality

- Inductor ← Current ← Magnetic field (B, H)

- Capacitor ← Charge ← Electric field (D, E)

- *But a changing electric field produces a magnetic field (Ampere's law)*

- *And a changing magnetic field produces an electric field (Faraday's law)*

Page 24 Copyright 2016, Sanjaya Maniktala

24

Creating Topologies

25

Creating a "topology"

- A) We need to suppress the uncontrolled voltage spike when we try to discharge an inductor--- and convert it into a *controlled induced voltage*

- B) To do that, we need to offer "current continuity" (perhaps with a diode?)

- C) We need to collect the energy of the "controlled induced voltage" and deliver it to an "output capacitor".

- D) We want to achieve "steady state" to call it a "topology", not an "event"

26

Let's use a freewheeling diode to control the spike

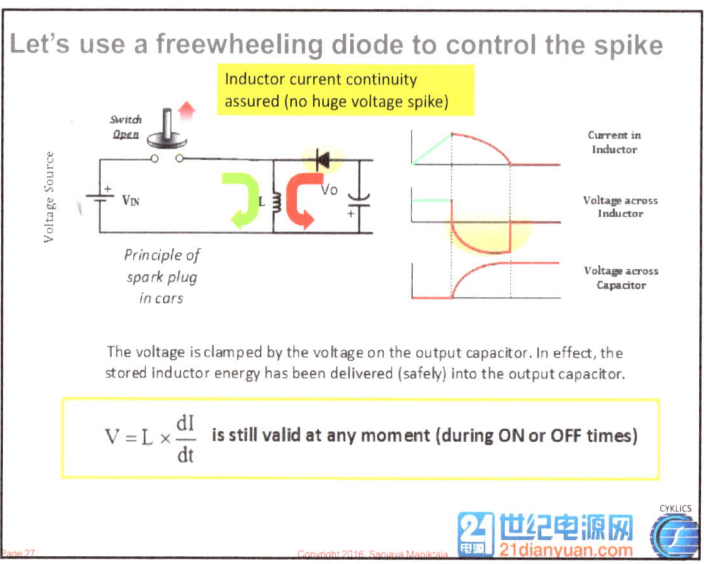

Inductor current continuity assured (no huge voltage spike)

Principle of spark plug in cars

The voltage is clamped by the voltage on the output capacitor. In effect, the stored inductor energy has been delivered (safely) into the output capacitor.

$$V = L \times \frac{dI}{dt}$$ **is still valid at any moment (during ON or OFF times)**

27

Let's apply a load, and switch **repetitively (with diode present and output capacitor)**

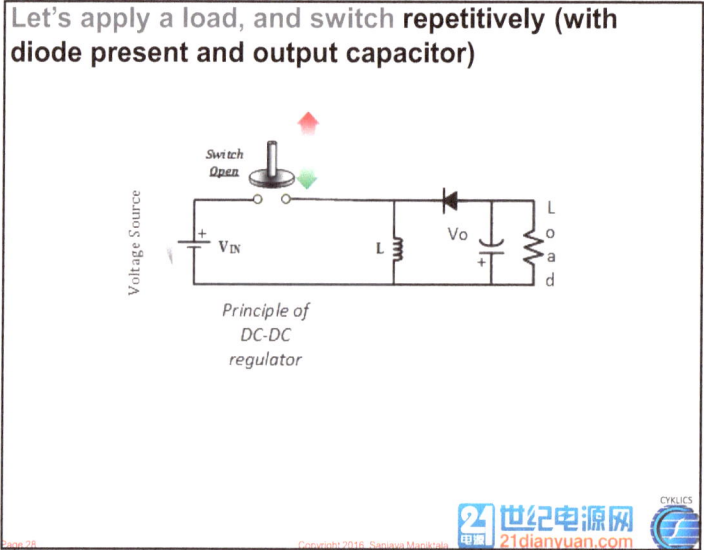

Principle of DC-DC regulator

28

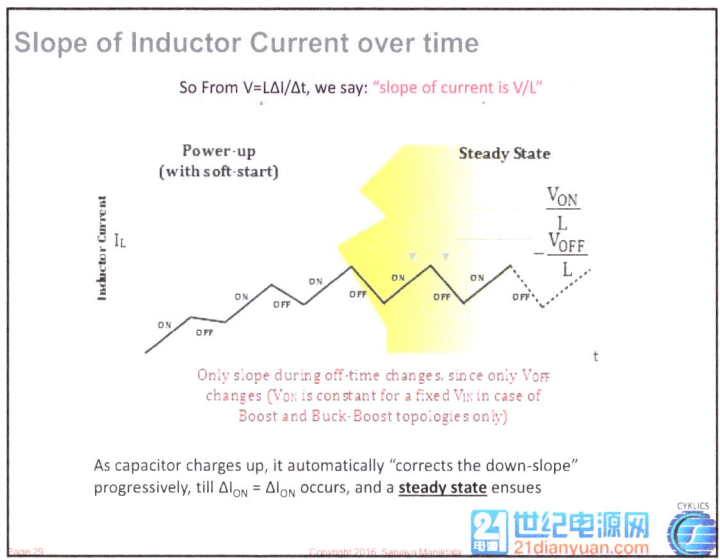

29

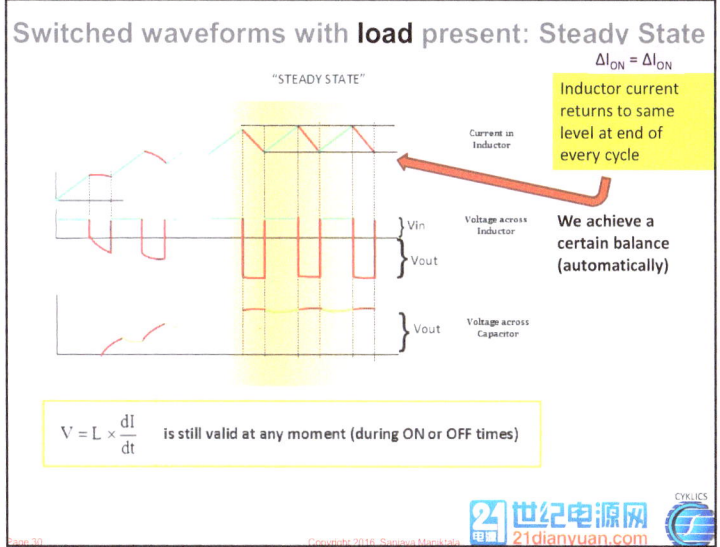

30

This is what we achieved (instead of disaster)

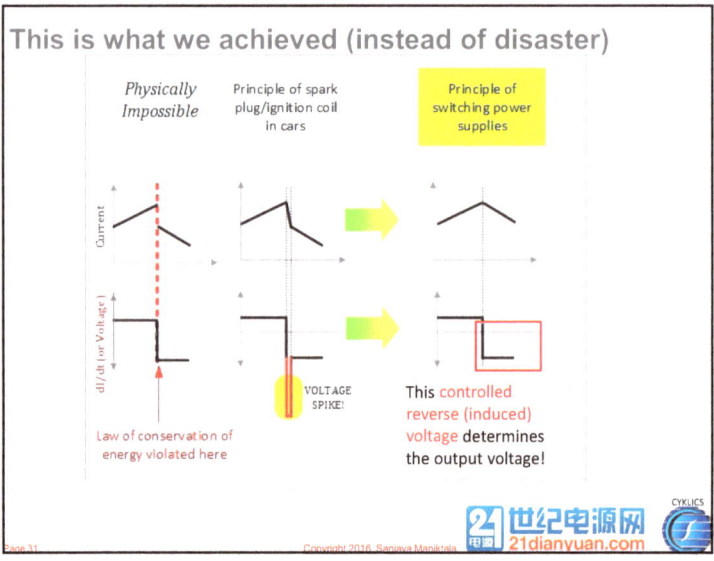

31

A (controlled, induced) voltage reversal must occur in <u>Steady State</u>

- To cause inductor current to slope down during OFF-time, to offset the rise during the ON-time, the inductor voltage **must reverse** (have an opposite/negative sign)

- If inductor voltage reversal never occurs, the "topology" is **not** a stable/valid topology! *(Don't rush to patent it!)*

- *The reverse voltage actually determines the output voltage. And since the reverse voltage depends on duty cycle, by varying the duty cycle, we can regulate the output voltage!*

32

Steady State and the <u>Volt-seconds Law</u>

- In any valid topology, this state **will** be achieved. Then:

- A) Inductor current returns to the <u>same value</u> at end of each cycle as it started off with
  - (which is why things repeat themselves exactly every cycle, unless line or load disturbances occur)
- B) This means the increase in current during the "ON time", must be exactly canceled by a decrease in current during the "OFF time"...So $| \Delta I_{ON} | = | \Delta I_{OFF}|$

- C) Based on $V = L \times \Delta I / \Delta t$, this means
$$L \Delta I = |V_{ON} \times t_{ON}| = |V_{OFF} \times t_{OFF}|$$

V_{ON} and V_{OFF} are the voltages across the inductor during the ON and OFF times respectively

33

The Volt-seconds Law (in Steady State)

- The applied "volt-seconds" during on-time is $V_{ON} \times t_{ON}$

- The applied "volt-seconds" during off-time is $V_{OFF} \times t_{OFF}$

Definition of steady state: The <u>net</u> applied volt-seconds during a switching cycle must be zero (otherwise current will **never** stabilize)

Positive volt-seconds causes a proportional increase in current
Negative volt-seconds causes a proportional decrease in current

For the increase in current to equal the decrease in current, the applied volt-seconds during ON time must be equal and opposite to the applied volt-seconds during the OFF time....(that is "steady state" by definition)

34

Volt-seconds "Et"

- It is the applied (constant) voltage multiplied by the duration it is applied for

- Often abbreviated as Et or ET or Vsec.....

$$\text{From } V = L\Delta I / \Delta t:$$
$$Et = L\Delta I$$

Note: "Et" dictates the AC component (swing) of inductor current it does not tell us any thing about the DC (center) value, which is determined by the applied load

35

Thought Process

ALBERT

How can we regulate the output?

36

Duty Cycle (Pulse Width) Modulation/Regulation

- Also called Pulse Width Modulation (PWM) is how we practically regulate the output to a certain desired **set point** (a fixed, desired value such as 3.3V or 5V or 12V etc)

- By changing the ON-time pulse width, we will change the applied "Et". *So the Et during the OFF-time will automatically change*, in an effort to restore steady state. But this changes the controlled reverse (induced) voltage and that changes the output rail.

37

How a change in D changes Vout

In both cases, $\Delta I_{ON} = \Delta I_{OFF}$ ultimately (automatically)

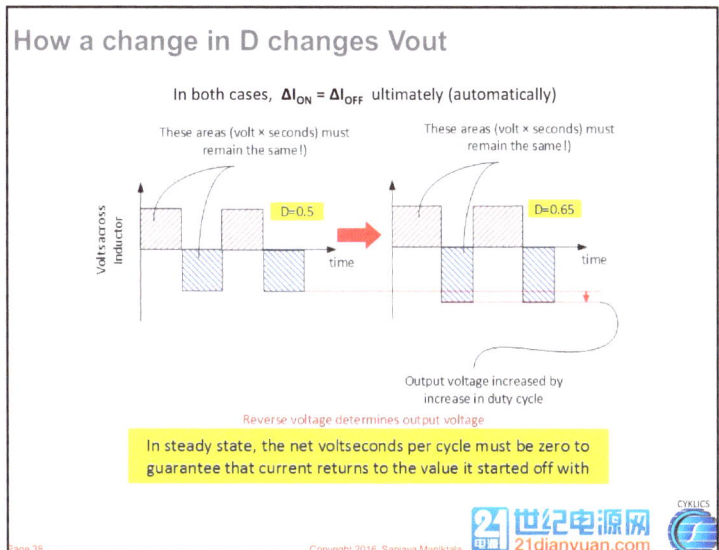

These areas (volt × seconds) must remain the same!)

These areas (volt × seconds) must remain the same!)

D=0.5

D=0.65

Volts across Inductor

time

time

Output voltage increased by increase in duty cycle

Reverse voltage determines output voltage

In steady state, the net voltseconds per cycle must be zero to guarantee that current returns to the value it started off with

38

Duty Cycle (Pulse Width) Modulation/Regulation

- By an active closed loop feedback system, we can maintain Vout irrespective of line and load variations....

- Typically, if input falls or load increases, the output will tend to fall, and the controller will automatically increase the duty cycle to correct the situation.

39

Pulse Width Modulation (Varying D to regulate Vout)

$$\text{Duty Cycle "D"} = \frac{\text{ton}}{\text{ton} + \text{toff}} = \frac{\text{ton}}{\text{T}}$$

$\text{where} \, T = 1/f$

$T = \text{time period}, f \text{ is the switching frequency}$

We can regulate the output voltage by changing D

This is called Pulse Width Modulation ("PWM")

In all topologies, if output voltage starts to fall, as it will either due to :

a) An increase in load, or b) a fall in input voltage,

we need to increase D to regulate properly (i.e. to bring back output rail to "setpoint")

40

Duty Cycle: Definition

- Duty cycle "D" is t_{ON}/T, where $T = 1/f$ (f is the switching frequency). So, using $V = L\, dI/dt$, in steady state we get:

(Ignoring signs)

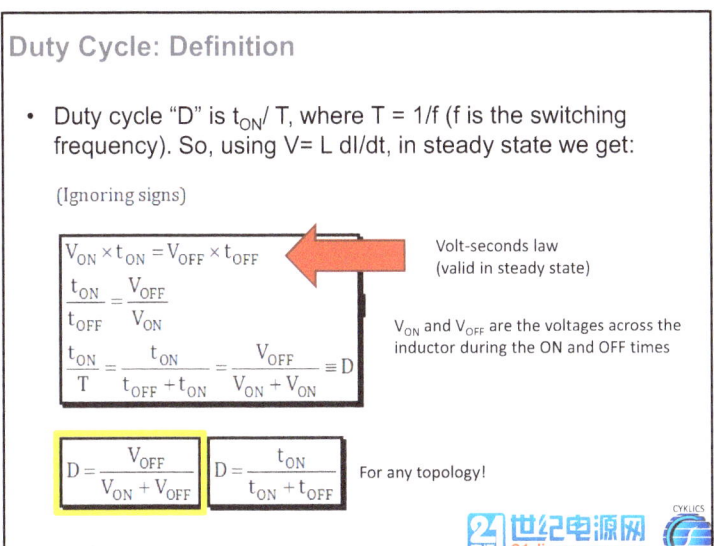

$$V_{ON} \times t_{ON} = V_{OFF} \times t_{OFF}$$

$$\frac{t_{ON}}{t_{OFF}} = \frac{V_{OFF}}{V_{ON}}$$

$$\frac{t_{ON}}{T} = \frac{t_{ON}}{t_{OFF} + t_{ON}} = \frac{V_{OFF}}{V_{ON} + V_{ON}} \equiv D$$

Volt-seconds law
(valid in steady state)

V_{ON} and V_{OFF} are the voltages across the inductor during the ON and OFF times

$$D = \frac{V_{OFF}}{V_{ON} + V_{OFF}} \qquad D = \frac{t_{ON}}{t_{ON} + t_{OFF}}$$

For any topology!

41

ON and OFF Voltages (across Inductor): Buck

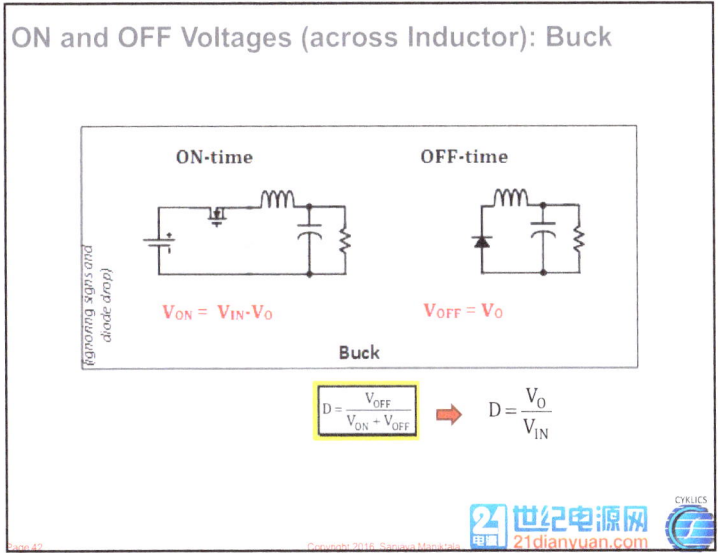

ON-time   OFF-time

(Ignoring signs and diode drop)

$V_{ON} = V_{IN} - V_O$ $V_{OFF} = V_O$

Buck

$$D = \frac{V_{OFF}}{V_{ON} + V_{OFF}} \implies D = \frac{V_O}{V_{IN}}$$

42

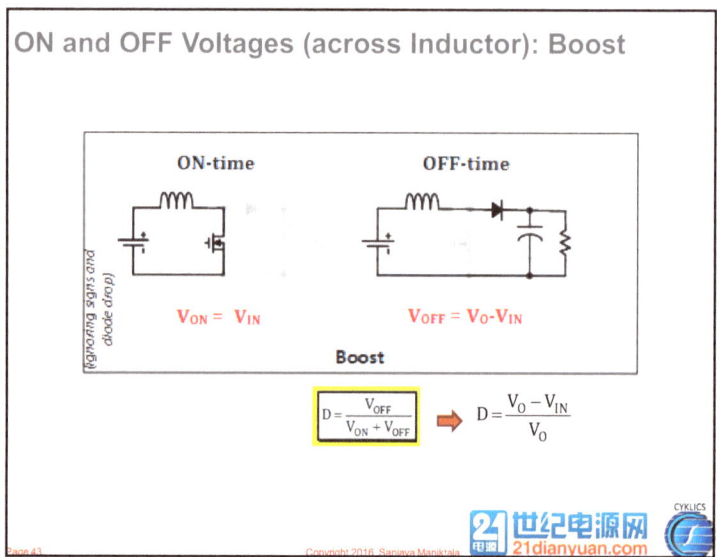

43

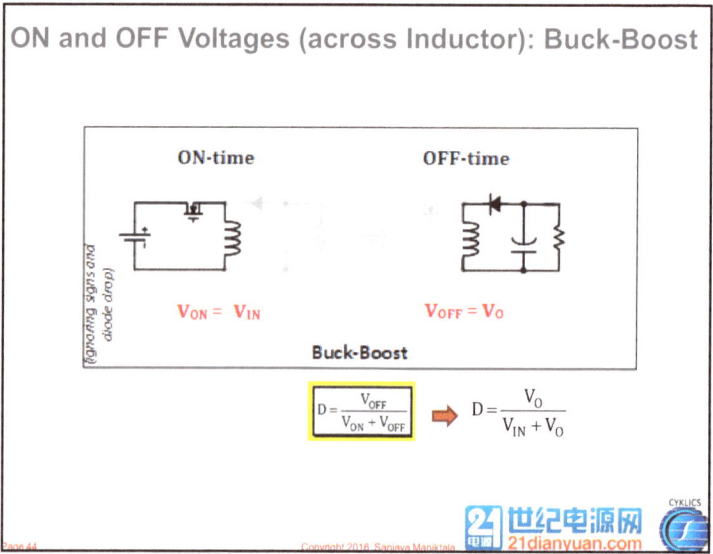

44

Basic DC-DC Design Equations

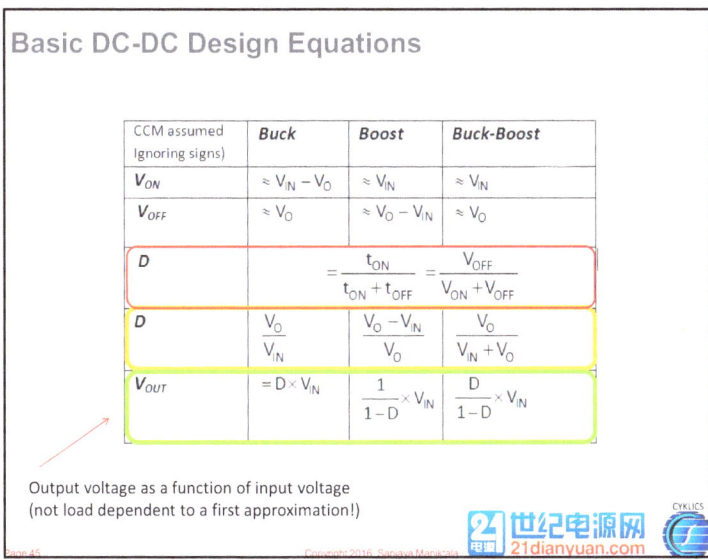

CCM assumed (Ignoring signs)	*Buck*	*Boost*	*Buck-Boost*
V_{ON}	$\approx V_{IN} - V_O$	$\approx V_{IN}$	$\approx V_{IN}$
V_{OFF}	$\approx V_O$	$\approx V_O - V_{IN}$	$\approx V_O$
D		$= \dfrac{t_{ON}}{t_{ON} + t_{OFF}} = \dfrac{V_{OFF}}{V_{ON} + V_{OFF}}$	
D	$\dfrac{V_O}{V_{IN}}$	$\dfrac{V_O - V_{IN}}{V_O}$	$\dfrac{V_O}{V_{IN} + V_O}$
V_{OUT}	$= D \times V_{IN}$	$\dfrac{1}{1-D} \times V_{IN}$	$\dfrac{D}{1-D} \times V_{IN}$

Output voltage as a function of input voltage
(not load dependent to a first approximation!)

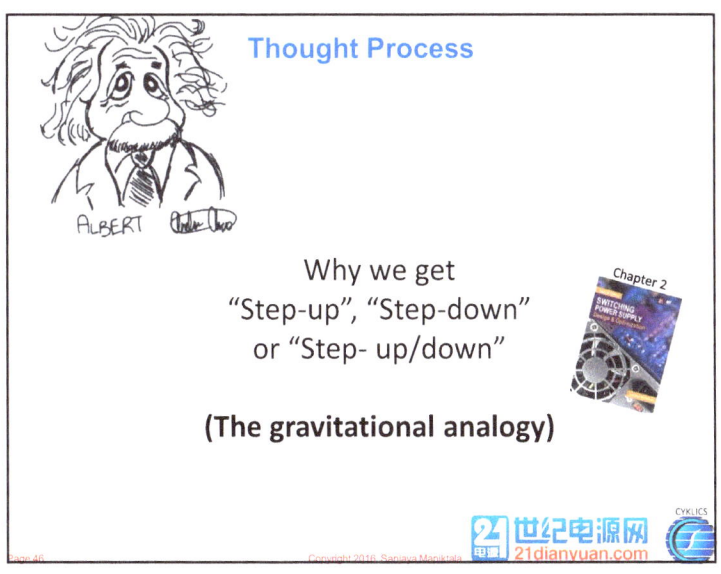

Thought Process

ALBERT

Why we get
"Step-up", "Step-down"
or "Step- up/down"

Chapter 2

(The gravitational analogy)

Why Buck topology only steps-down

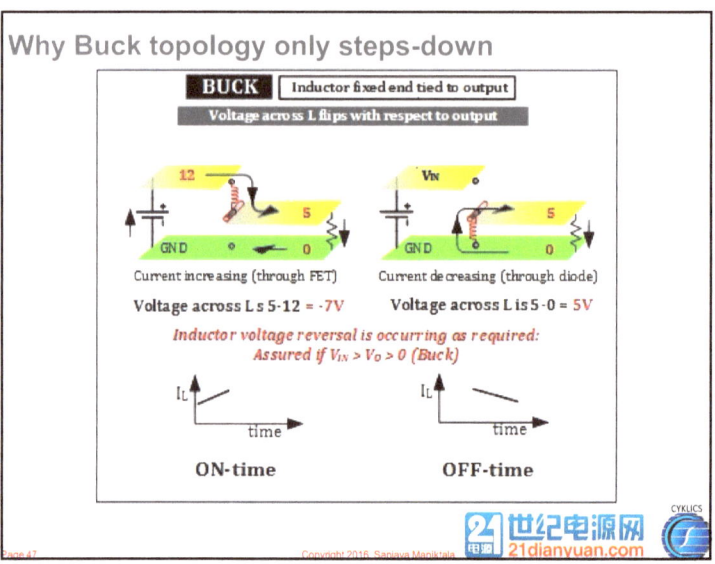

47

Why Boost topology only steps-up

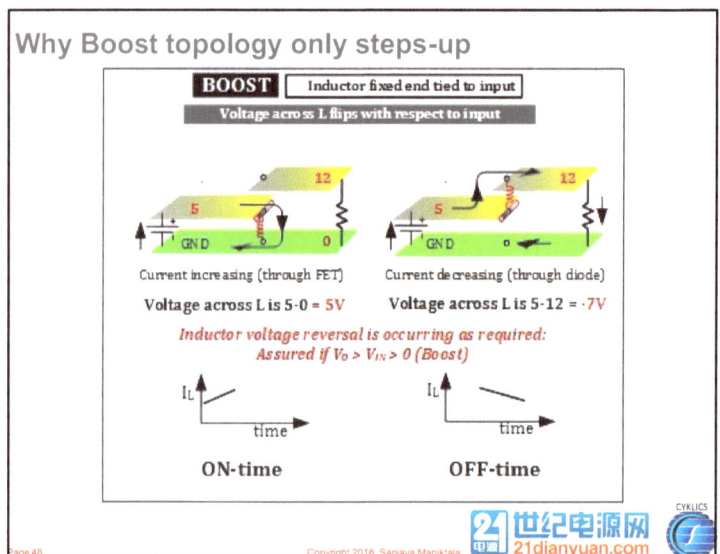

48

Why Buck-Boost topology steps up or down

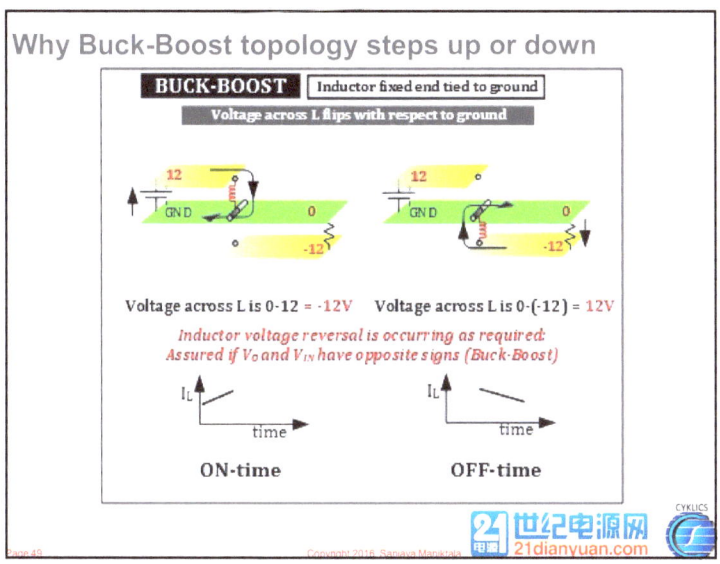

49

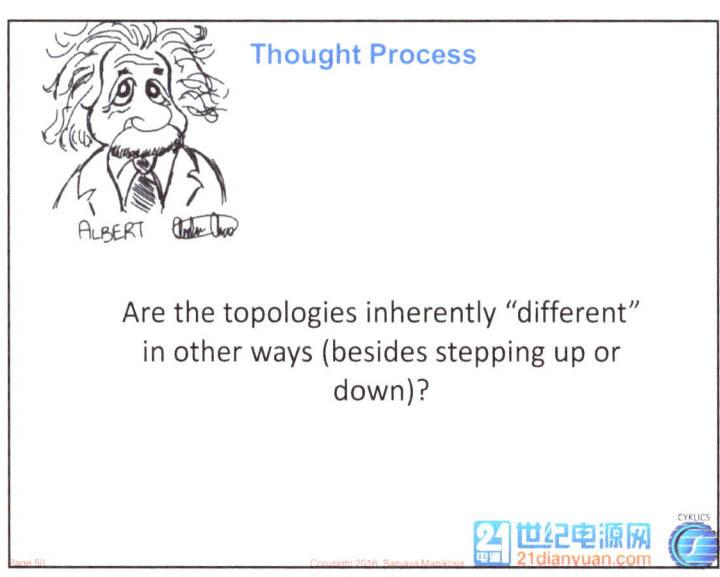

50

Why the inherent efficiency of the Buck-Boost is <u>low</u>

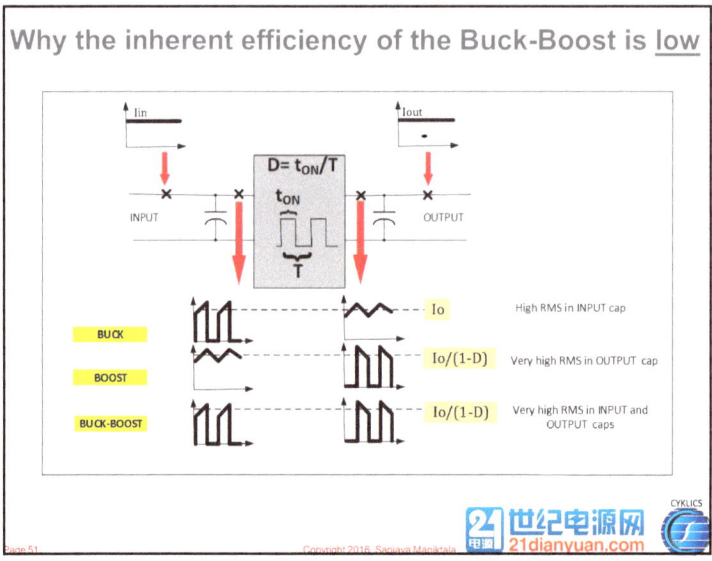

51

Energy Transfer Differences

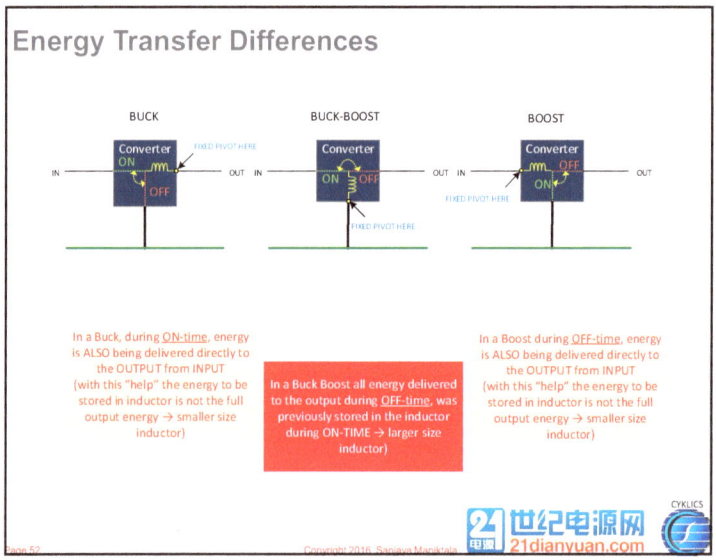

52

Why the size of the Buck-Boost is *inherently* big

- Comparing Core Sizes (energy packet size per cycle)

$$\Delta\varepsilon = \frac{P_{IN}}{f} \times (1 - D) \quad (\text{buck})$$

$$\Delta\varepsilon = \frac{P_{IN}}{f} \times D \quad (\text{boost})$$

$$\Delta\varepsilon = \frac{P_{IN}}{f} \quad (\text{buck}-\text{boost})$$

The peak energy storage is:

$$\varepsilon_{PEAK} = \frac{\Delta\varepsilon}{8} \times \left[r \times \left(\frac{2}{r} + 1\right)^2 \right]$$

"r" is the **current ripple ratio** which I introduced in 2000 at National Semi. I recommended: set it to **0.4** for optimum results (any topology, any application)

So the core size of the Buck-Boost topology is the highest of all topologies, for a given power level (and same *r*).

53

"r" current ripple ratio

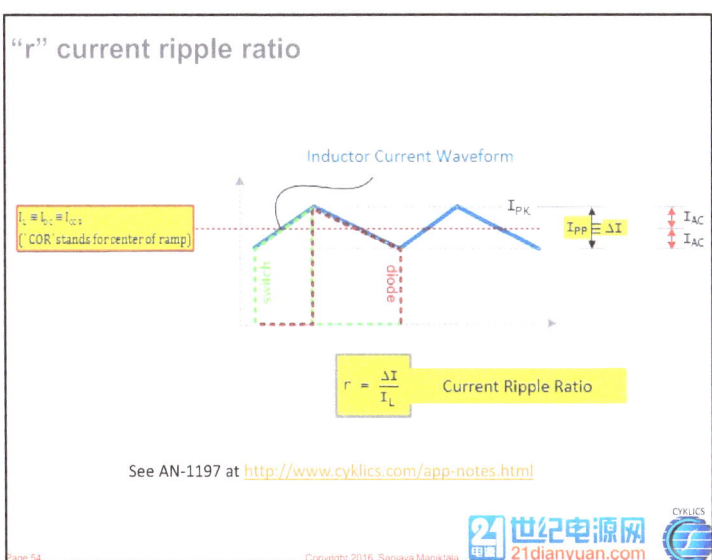

Inductor Current Waveform

$$r = \frac{\Delta I}{I_L} \quad \text{Current Ripple Ratio}$$

See AN-1197 at http://www.cyklics.com/app-notes.html

54

Why the size of the Buck-Boost is *inherently* big

- *Another way to look at this is:*
- In a Buck-Boost, all the energy delivered ("Pin/f" per cycle) was previously stored in the inductor. During the ON-time energy is completely stored in the inductor alone, during the OFF-time, this alone is extracted and delivered to the output

- In contrast, in a Buck, energy is delivered to the output during both the ON and OFF times.
- In a Boost, energy is drawn from the input during both the ON and OFF times

So, in a Buck or Boost, only a fraction of the energy delivered to the output needs to be stored in the inductor...not in a Buck-Boost!

55

An Odd Behavior of Boost and Buck-Boost

- If input falls, or output load increases suddenly, a closed loop system will try to correct it by increasing ON-time (i.e. D).

- But that leaves even less time for the stored energy to be delivered to the output since in both topologies, unlike a Buck, energy is delivered to output ONLY during OFF-time.

- So the output falls further momentarily, instead of increasing.

- Mathematically this leads to the "right half plane zero" (RHPZ) issue in control loop theory

56

An Odd Behavior of Boost and Buck-Boost

- The RHPZ is almost impossible to deal with, and the only practical solution is to slow down the correction speed ("lower the crossover frequency of the Gain plot").

- Which also means that the response time of a Boost and a Buck-Boost has typically to be set ~ 5-10 times slower than a Buck.

- Yes, if the Boost and Buck-Boost are operated in "Discontinuous Conduction Mode" ("DCM") or in Boundary Conduction mode ("BCM"), the RHPZ moves to a very high frequency and can be effectively ignored. So we can set faster response time in that case!

Page 57   Copyright 2016, Sanjaya Maniktala

57

CCM to DCM

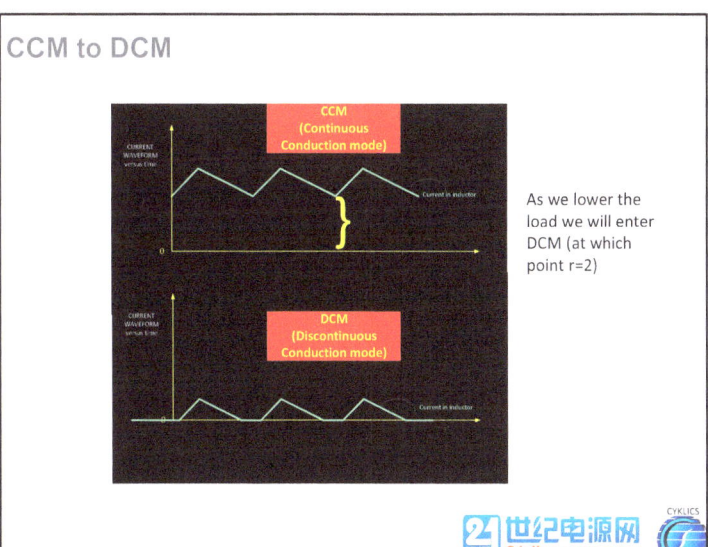

As we lower the load we will enter DCM (at which point r=2)

Page 58   Copyright 2016, Sanjaya Maniktala

58

29

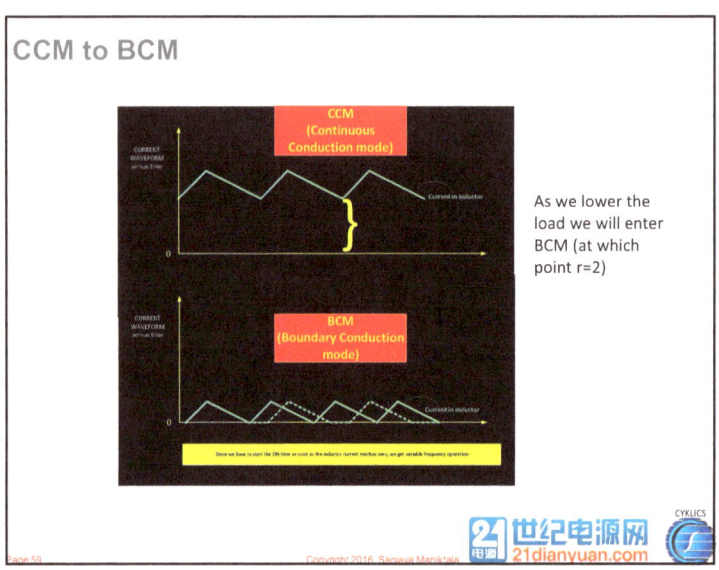

59

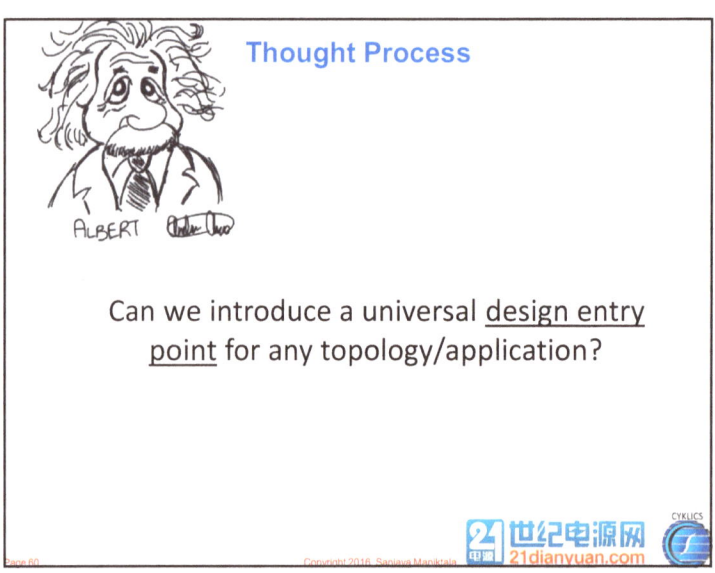

60

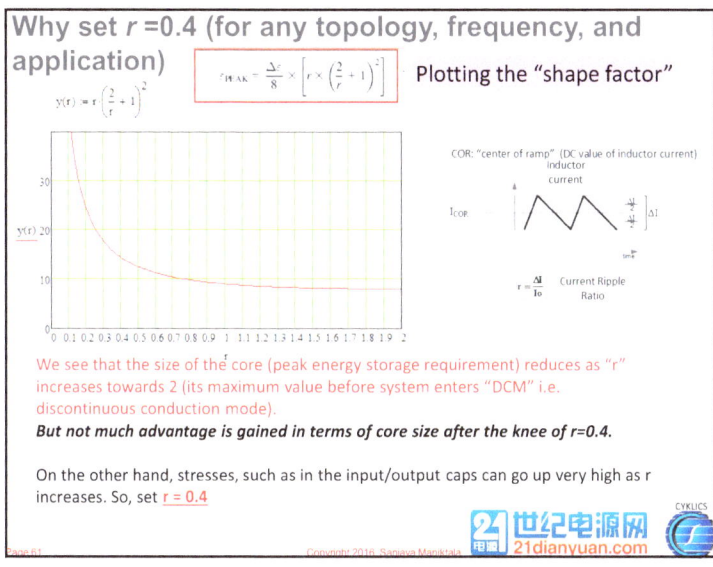

Why set r =0.4 (for any topology, frequency, and application)

$$z_{PEAK} = \frac{\Delta z}{8} \times \left[r \times \left(\frac{2}{r} + 1 \right)^2 \right]$$

$$y(r) = r \left(\frac{2}{r} + 1 \right)^2$$

Plotting the "shape factor"

We see that the size of the core (peak energy storage requirement) reduces as "r" increases towards 2 (its maximum value before system enters "DCM" i.e. discontinuous conduction mode).

But not much advantage is gained in terms of core size after the knee of r=0.4.

On the other hand, stresses, such as in the input/output caps can go up very high as r increases. So, set r = 0.4

61

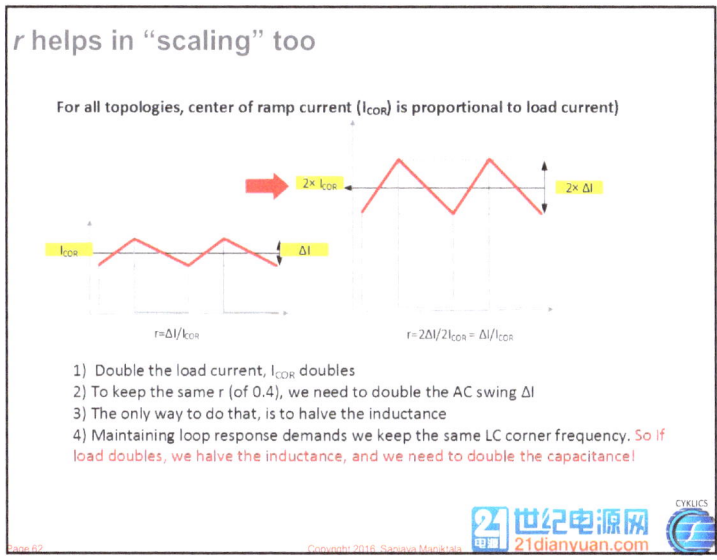

r helps in "scaling" too

For all topologies, center of ramp current (I_{COR}) is proportional to load current)

$r = \Delta I / I_{COR}$

$r = 2\Delta I / 2I_{COR} = \Delta I / I_{COR}$

1) Double the load current, I_{COR} doubles
2) To keep the same r (of 0.4), we need to double the AC swing ΔI
3) The only way to do that, is to halve the inductance
4) Maintaining loop response demands we keep the same LC corner frequency. So if load doubles, we halve the inductance, and we need to double the capacitance!

62

COR value for different topologies

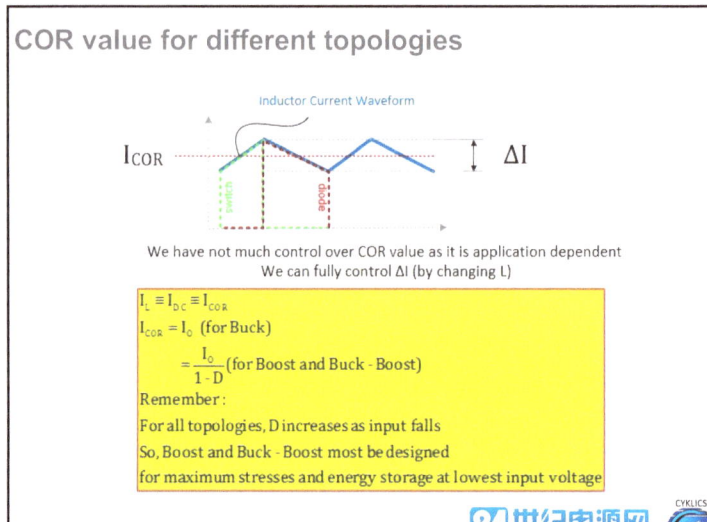

Inductor Current Waveform

I_{COR} ΔI

We have not much control over COR value as it is application dependent
We can fully control ΔI (by changing L)

$I_L \equiv I_{DC} \equiv I_{COR}$
$I_{COR} = I_O$ (for Buck)
$\quad = \dfrac{I_O}{1-D}$ (for Boost and Buck-Boost)
Remember :
For all topologies, D increases as input falls
So, Boost and Buck-Boost most be designed
for maximum stresses and energy storage at lowest input voltage

63

Scaling Laws we "know" intuitively

- All else unchanged, if we double the load, we need to halve the inductance <u>and</u> double the capacitance! -- **"Power Scaling Law"**

The reason we go to higher frequencies is to reduce the size of both inductors and capacitors! So…

- If we double switching frequency, we halve the inductance and halve the capacitance. -- **"Frequency Scaling Law"**

In all cases above, the set r of 0.4 is unchanged. It is the basis for the above scaling laws!

64

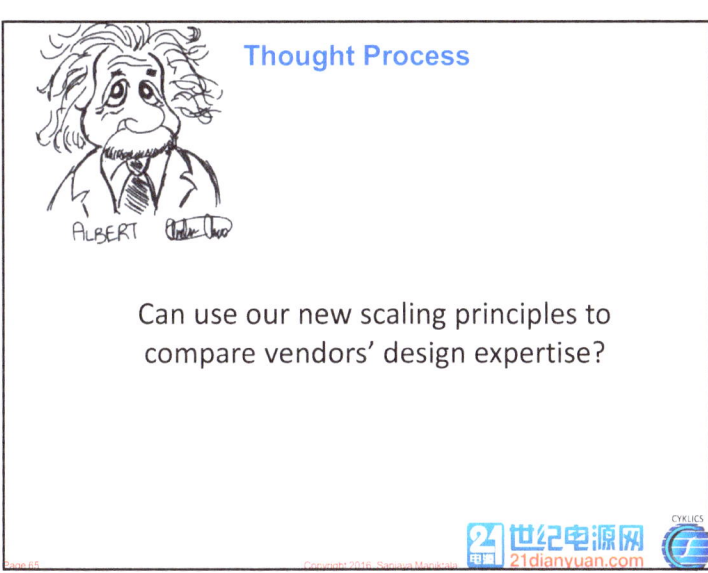

65

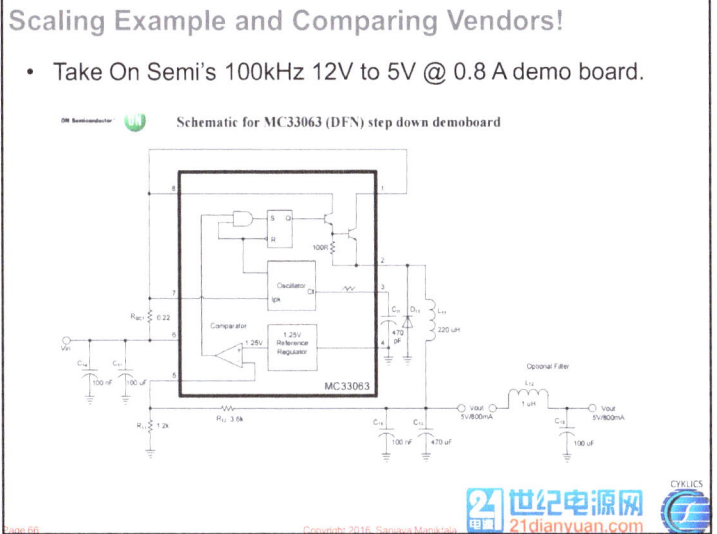

66

Using Design Equation Table (Chapter 20) of A-Z/2e

- For the Buck, Boost and Buck-Boost this tells you the desired inductance

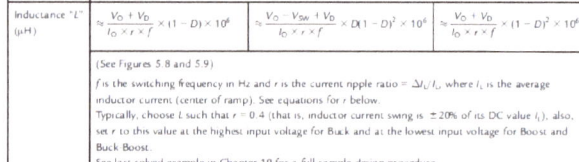

Inductance "L" (µH)	$\approx \frac{V_O + V_D}{I_O \times r \times f} \times (1 - D) \times 10^6$	$\approx \frac{V_O - V_{SW} + V_D}{I_O \times r \times f} \times D(1 - D)^2 \times 10^6$	$\approx \frac{V_O + V_D}{I_O \times r \times f} \times (1 - D)^2 \times 10^6$

(See Figures 5.8 and 5.9)

f is the switching frequency in Hz and r is the current ripple ratio $= \Delta I_L / I_L$, where I_L is the average inductor current (center of ramp). See equations for r below.

Typically, choose L such that $r = 0.4$ (that is, inductor current swing is $\pm 20\%$ of its DC value I_L), also, set r to this value at the highest input voltage for Buck and at the lowest input voltage for Boost and Buck-Boost.

See last solved example in Chapter 19 for a full sample design procedure.

$$V_{in} = 12 \qquad V_O = 5 \qquad f = 100 \cdot 10^3 \qquad I_O = 0.8 \qquad r = 0.4$$

$$D = \frac{V_O}{V_{in}}$$

$$L := \frac{V_O}{I_O \cdot r \cdot f} (1 - D) \cdot 10^6$$

$$L = 91.146 \quad (uH)$$

This tells us that we need a ~ 100uH inductor, based on $r = 0.4$

67

Scaling applied

- The Evaluation board however uses L=220uH, for an r of 0.2 (too low). Also Cin = 100uF, Cout = 470uF. But let's accept these values for now.

- Let's apply scaling to go to a freq of 150kHz (1.5 times higher). The components would scale to 220uH/1.5=150uH and Cin=100uF/1.5=66uF, Cout=470uF/1.5=313uF.

- Let's apply scaling to go to a load of 3A (factor of 3A/0.8A=3.75). So inductance will *decrease* to 150/3.75=40uH. Similarly Cin will *increase* to 66 x 3.75=248uF, Cout will *increase* to 313 x 3.75 = 1174uF.

68

On-Semi scaled to 3A/150kHz and compared with LM2586 150kHz/3A Eval Board

- The LM 2596 board is

Typical Application
(Fixed Output Voltage Versions)

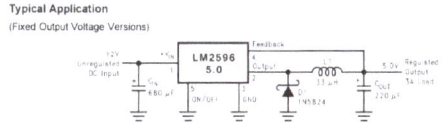

LM2596 uses (3A/150kHz):
L = 33uH
Cin=680uF
Cout=220uF

The LC break frequency based on 1/2πLCout=1.8kHz, which is 1.2% of the switching frequency (150kHz). **Typical values for the LC pole are 1-2% of fsw**, so this is correct.

On-Semi board scaled to 3A/150kHz:
L = 40uH
Cin=248uF
Cout=1174uF

The LC break frequency based on 1/2πLCout is just 0.5% of the switching frequency (150kHz). **Typical values for the LC pole are 1-2% of fsw**, so this is too low.

69

Analysis and Comparison of Designs

1) The correct value of inductance (for *r* =0.4) should be around 22uH (but the current limits of both devices are apparently too low, and the *r had to be decreased to 0.2 to avoid hitting current limit…discussed later*)

2) *LM2596 uses a very large input capacitor, indicating noise sensitivity, which On-Semi avoided by a 0.1uF ceramic decoupling cap in parallel to the bulk cap (good idea)*

3) *On-Semi uses an excessively large output cap, which will lead to* slow loop response *(poorer correction)…LM2596 has the right LC break frequency (**between 1-2% of switching frequency**).*

The secret to fast loop response is to have small inductors and capacitors if possible.

70

35

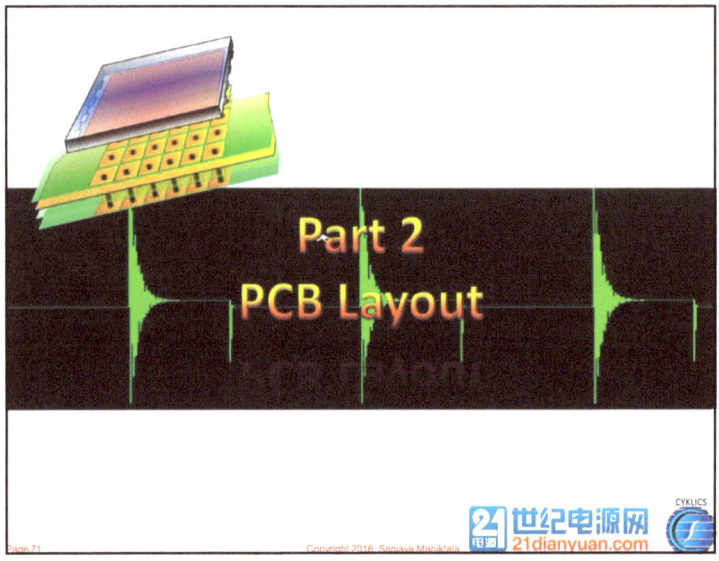

71

The main source of problems in switching power

In design, testing, certification, reliability, PCB layout etc…:

• They are almost always related to switching "very fast"

But what does "switching fast" really mean?
And why is that such a problem?

72

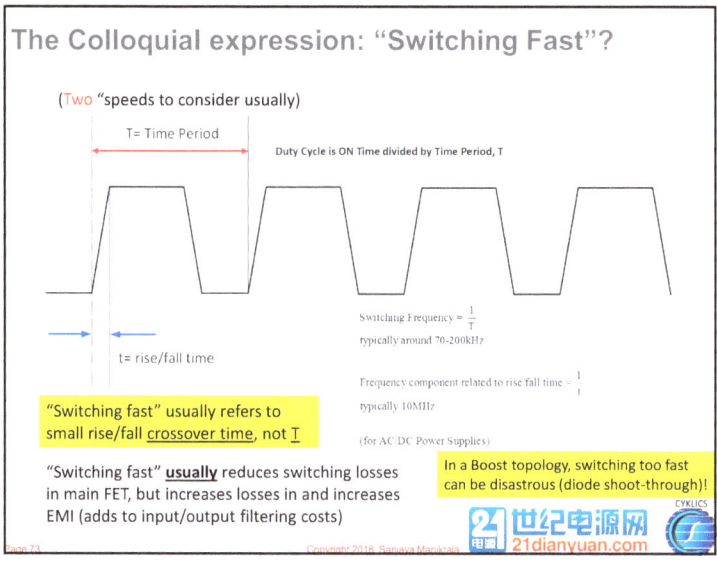

The Colloquial expression: "Switching Fast"?

(Two "speeds to consider usually)

T= Time Period

Duty Cycle is ON Time divided by Time Period, T

t= rise/fall time

Switching Frequency = $\frac{1}{T}$
typically around 70-200kHz

Frequency component related to rise/fall time = $\frac{1}{t}$
typically 10MHz

(for AC/DC Power Supplies)

"Switching fast" usually refers to small rise/fall crossover time, not T

"Switching fast" **usually** reduces switching losses in main FET, but increases losses in and increases EMI (adds to input/output filtering costs)

In a Boost topology, switching too fast can be disastrous (diode shoot-through)!

73

LESSONS LEARNED OVER TIME

Bay Area~1998

74

A High-power AC/DC Power Supply

- Extremely stubborn EMI (electromagnetic interference) was detected and traced to the small 5-10W "standby power supply" (Topswitch-II).... it switched "*extremely fast*".

- Its switching frequency was only 100kHz, but its <u>switch transition time</u> was very small (~ <30 ns all the way from 0 to hundred of volts). Also, since the FET was internal, there was no way to "slow" down the FET by increasing the gate resistor for example.

- **Solution:** Bigger/more expensive front-end EMI filters. The TOPSWITCH was not cheap after all! A two layer board had to be used too.

75

"Slowing the FET" (<u>increasing</u> crossover time) can lower EMI filtering Cost and Size

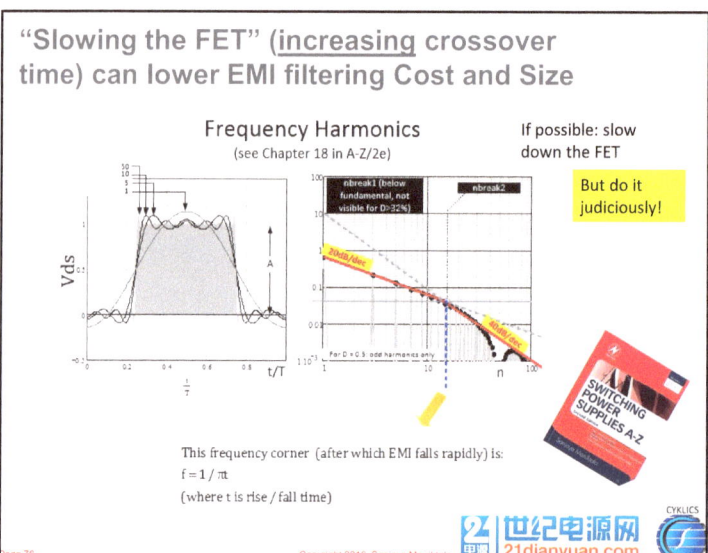

Frequency Harmonics
(see Chapter 18 in A-Z/2e)

If possible: slow down the FET

But do it judiciously!

This frequency corner (after which EMI falls rapidly) is:

$f = 1 / \pi t$

(where t is rise / fall time)

76

Even half-inch of PCB trace has a high impedance

- It is often said that an "*inch of PCB trace has an inductance of 20 nH*". Suppose we have just <u>half an inch of trace, i.e. 10nH.</u>

- If the rise/fall time is 20 ns (typical in a DC-DC converter), then the associated "break frequency" (of harmonics spectrum) can be shown to be about *1/3 rd the transition time*, i.e. 1/ (3 x 20n) = 16 MHz. (High EMI!)

- At this frequency, the impedance of 10 nH is from Z= L x 2π x f = 1 ohm…which is <u>very significant</u> considering it was "hidden" (a *parasitic*). It can impede any desired current swing, causing *voltage spikes*.

77

Half-inch of PCB trace produces voltage spikes

- Using V=LdI/dt, (the "inductor equation"), if we switch **5A** through ½ inch of PCB trace in 20ns, we get a momentary voltage of 2.5V across it (high voltage "spike").

- This has the capacity to destroy a low voltage FET directly, but more likely by simply turning it ON at the wrong moment, causing destruction through an "*unintended sequence of events*" (such as "shoot-through).

- Worse, in the case of *integrated switchers*, no pin is typically allowed to go about *0.3V below ground* to protect its internal ESD (electrostatic discharge) structures…and this 2.5V spike can kill/weaken the switcher directly, or make it "misbehave" (performance issues), or cause destruction through another *unintended sequence of events*.

78

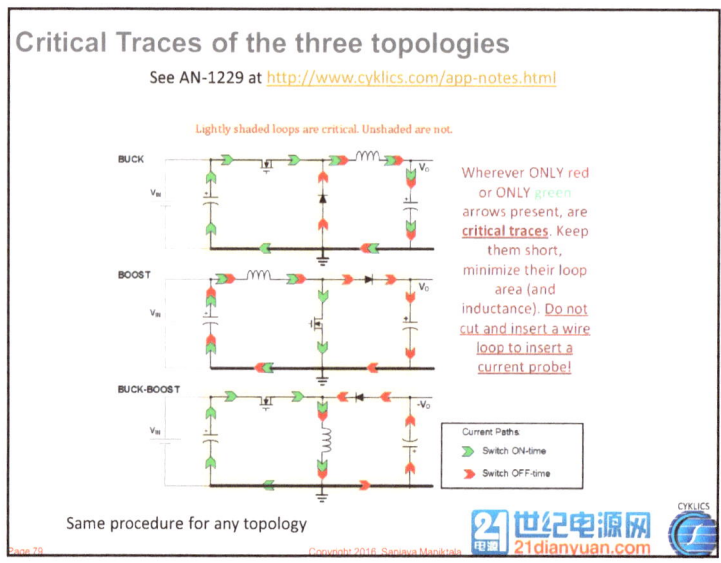

Critical Traces of the three topologies

See AN-1229 at http://www.cyklics.com/app-notes.html

Lightly shaded loops are critical. Unshaded are not.

BUCK

BOOST

BUCK-BOOST

Wherever ONLY red or ONLY green arrows present, are **critical traces**. Keep them short, minimize their loop area (and inductance). Do not cut and insert a wire loop to insert a current probe!

Current Paths:
- Switch ON-time
- Switch OFF-time

Same procedure for any topology

79

How to reduce PCB trace inductances!

- A) Increasing trace thickness hardly helps….because of self-inductance issues, you need to *typically increase the width 10 times to cause PCB trace inductance to halve*.

- B) The better option is to decrease the length…here halving the inductance can be accomplished by just halving the length…. But we don't have to do this for *all* traces, only those "critical traces": *which are expected to see the full current swing during each switch transition*.

- The most common way nowadays is to provide a nice ground plane under all traces (need a >2-layer PCB). This automatically causes the return current of any critical trace, to flow directly under the trace, producing almost full cancellation of the its magnetic field, thus reducing its inductance, without having to reduce its length.

80

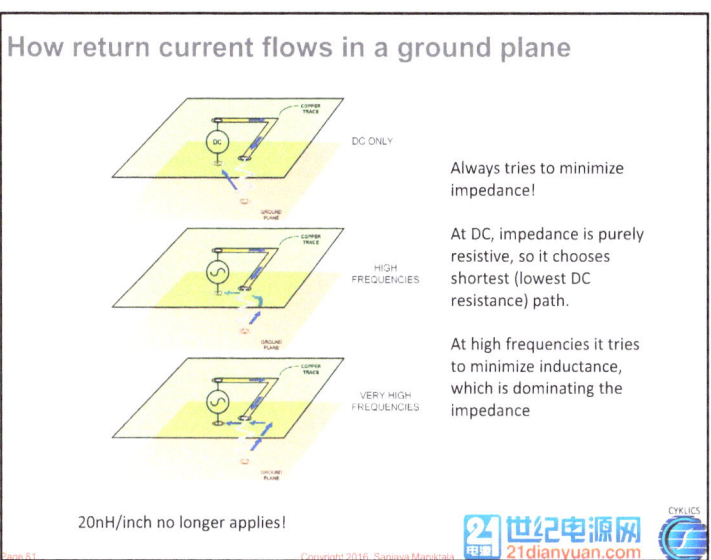

81

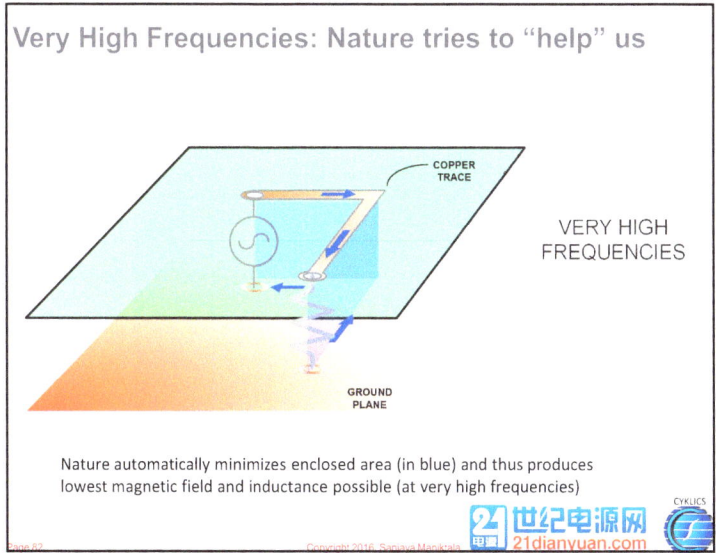

82

If we let it!

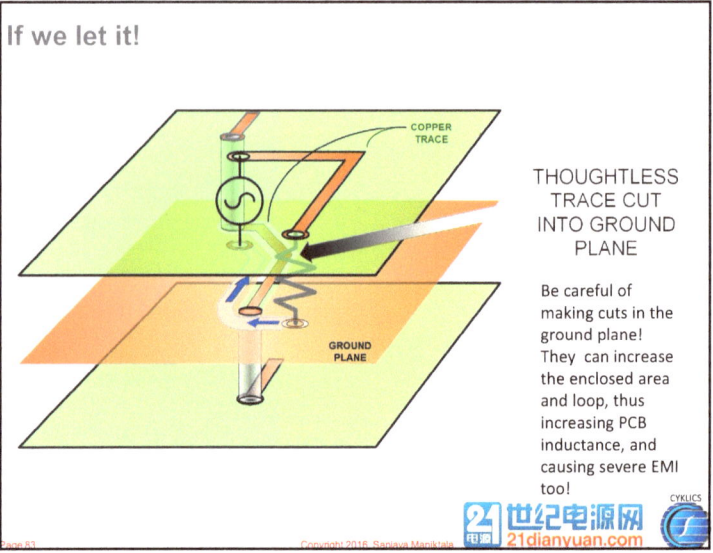

COPPER
TRACE

THOUGHTLESS
TRACE CUT
INTO GROUND
PLANE

Be careful of
making cuts in the
ground plane!
They can increase
the enclosed area
and loop, thus
increasing PCB
inductance, and
causing severe EMI
too!

GROUND
PLANE

83

Ensuring the Control IC can do its job well

- The spikes induced by poor layout can cause the control IC/switcher to malfunction, because it needs a "clean environment to do its job well".

- But the IC itself can create its own problems because it can demand sudden bursts of current as part of its own functioning. This can create noise on the supply rails of the IC, causing it to malfunction.

- So we need some small bulk capacitance (typically 1 to 4.7 µF) at the input of the IC, perhaps an inch o two away, but and also a 0.1 µF ceramic capacitance very close to the IC, to provide "decoupling/bypass" (its ability to reject noise and keep the supply rail to the IC clean).

- In some cases, such as a Buck topology, the IC's supply rail is the same as the input rail of the converter (Vin) so the IC supply capacitances mentioned above can be shared by the capacitors of the converter itself ("Cin").

84

A Real Capacitor Model has ESR and ESL

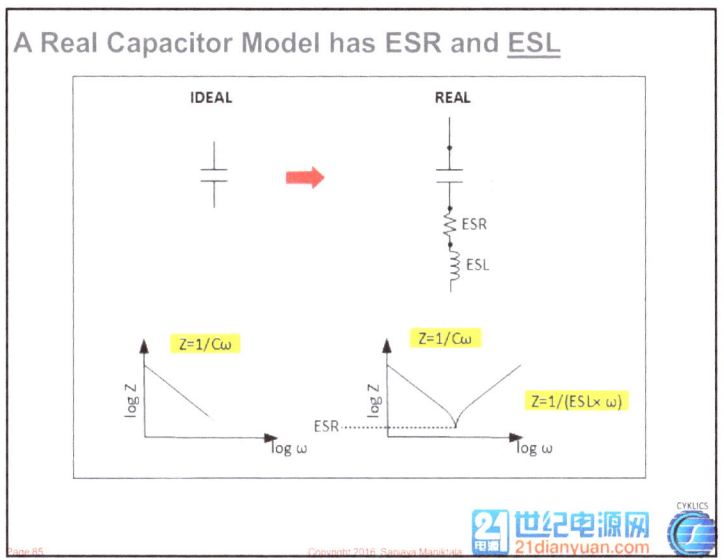

85

1uF ceramic capacitors compared

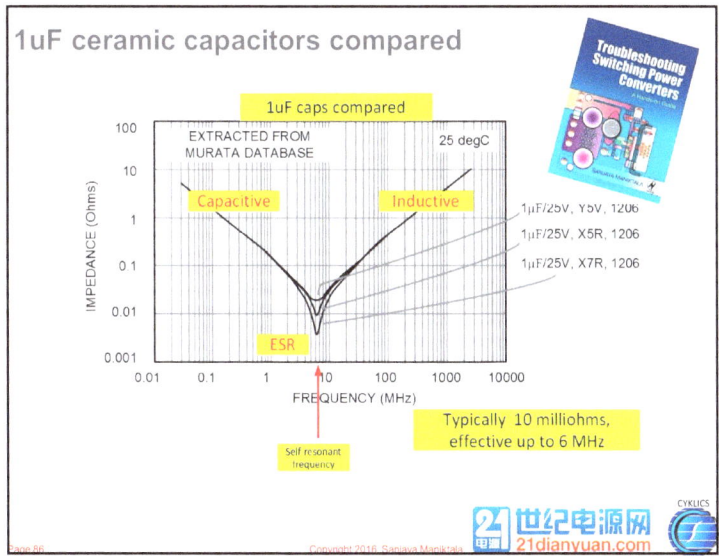

86

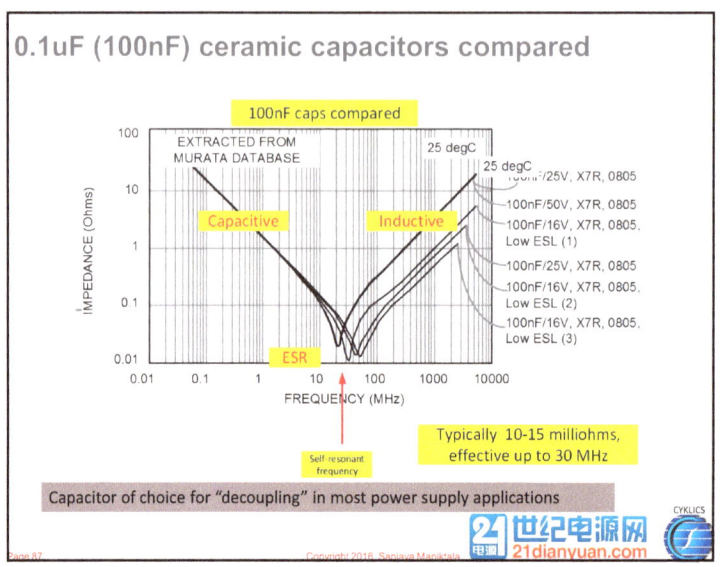

87

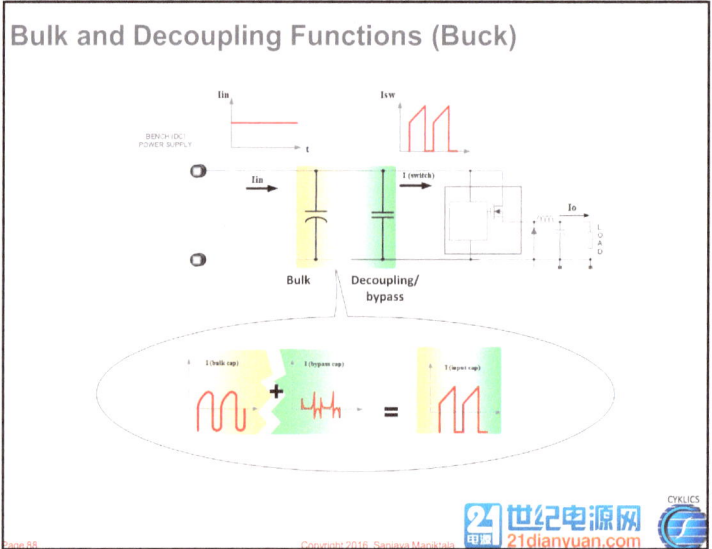

88

Use the 100nF capacitor on input and supply rails

- Every switcher/control IC needs a clean supply to operate properly. Put a 100nF cap very close to the IC supply and ground pins (<u>do not go through vias</u> from other side of board)

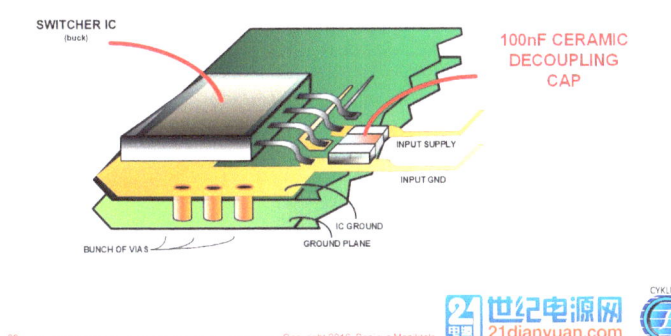

89

<u>Paralleling</u> ceramic capacitors may <u>not</u> help

- If of the same type and value, and we put "n" caps in parallel

$$C \Rightarrow nC$$

$$ESR \Rightarrow ESR/n$$

$$ESL \Rightarrow ESL/n$$

$$f_{resonance} \Rightarrow f_{resonance}$$

The self-resonant frequency does not change!

Certainly do not try to parallel capacitors to improve frequency response. Parallel it to lower ESR and thus **increase ripple current capability** (reduce I²R loss in output caps)

90

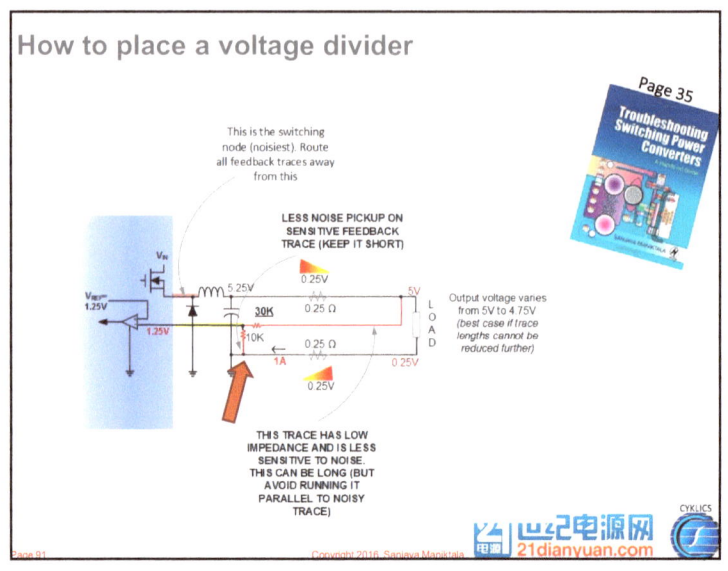

91

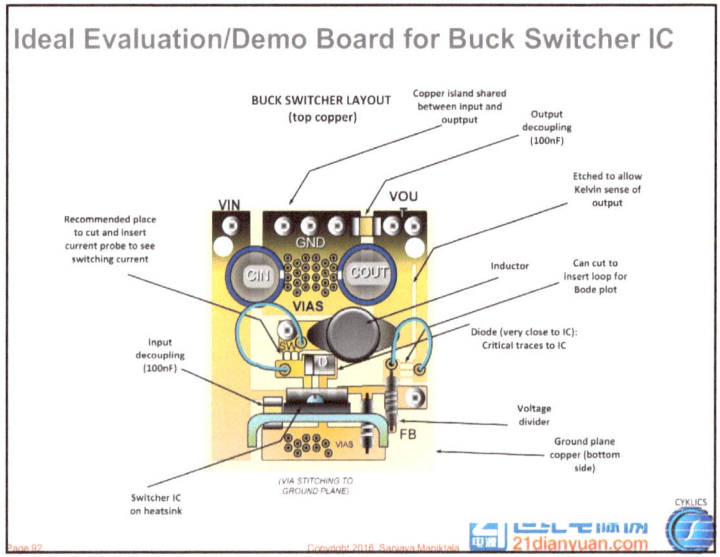

92

93

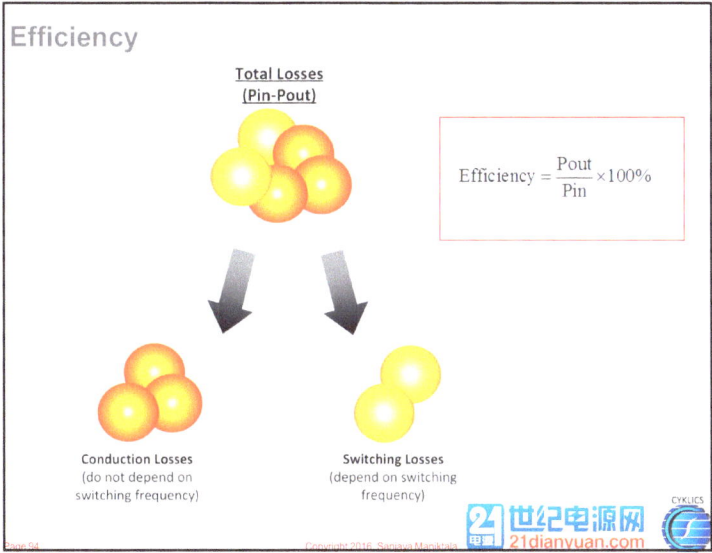

94

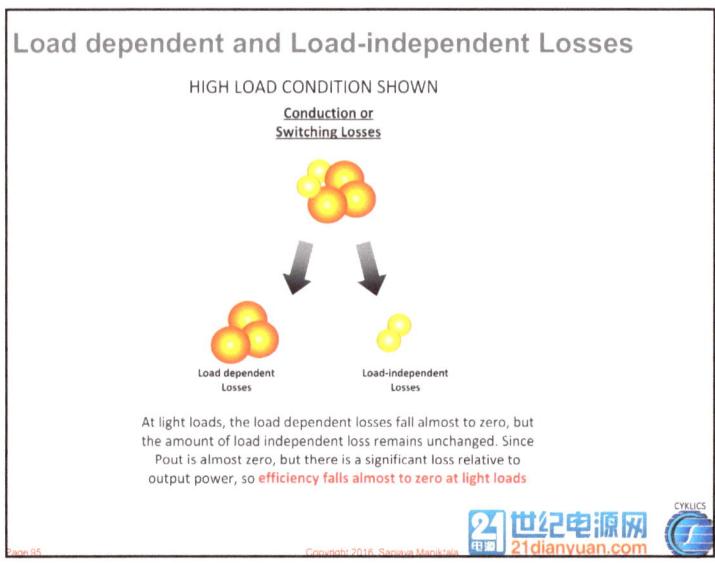

95

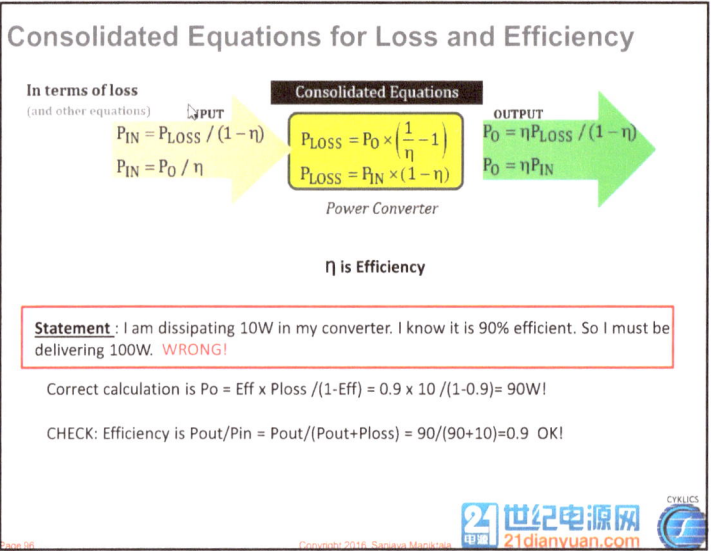

96

V-I Switching Waveforms

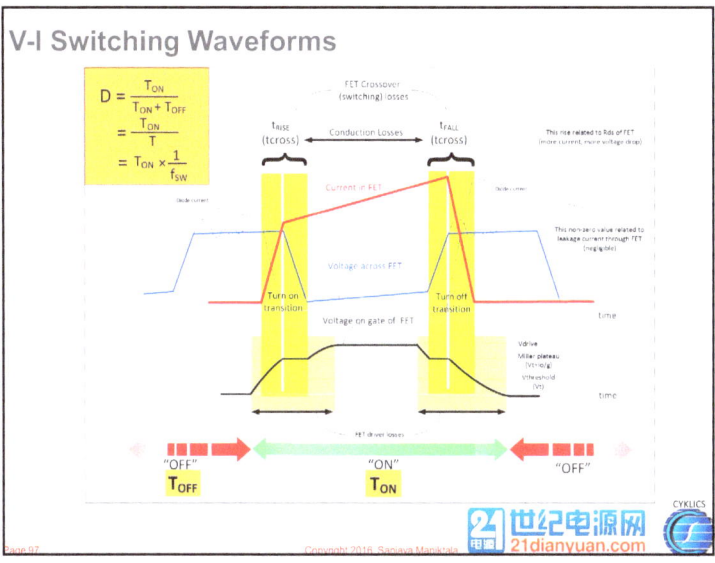

97

Switching and Conduction Losses

All losses take place in resistances, not in pure reactances!

- Key **Conduction Loss** Terms:
- A) "Rds" of switching FETs
- B) "DCR" of inductor/transformer
- C) "ESR" of input/output capacitors
- D) Quiescent Current (← dominates at light loads)

- Key **Switching Loss** Terms
- A) Depend on <u>crossover</u> (overlap) of voltage and current waveforms during switch transition
- B) FET driver dissipation (← dominates at light loads)
- C) "AC" (skin depth/core/proximity) losses"

98

Salient Points to Remember

- During turn-on transition, FET voltage does not even start to fall till FET current has reached its final value! (Because the freewheeling diode must continue to remain forward-biased till the FET has taken up the *entire* diode current)

- During turn-off transition, FET current does not even start to fall till FET voltage has reached its final value (Because the freewheeling diode must become fully forward-biased to even start taking up any current from the FET)

99

Salient Points to Remember

- Conduction Loss mainly depends on Rds of FET (its resistance when fully "ON")…does not depend on frequency
  (though I had to model in a concept called "dynamic/time-dependent Rds" to fully explain data and losses inn the TOPSWITCH while writing AN-26 and AN-29 at www.power.com

- Switching Loss depends on switching frequecy and also on "tcross". To lower it you need to "switch slowly" and also "switch fast"! *Well what is it?*

  To lower switching losses, lower the basic switching frequency ("switch slow"), and also reduce crossover time ("switch fast") --- but watch out for high EMI and PCB trace layout issues!

100

Switching and Conduction Losses

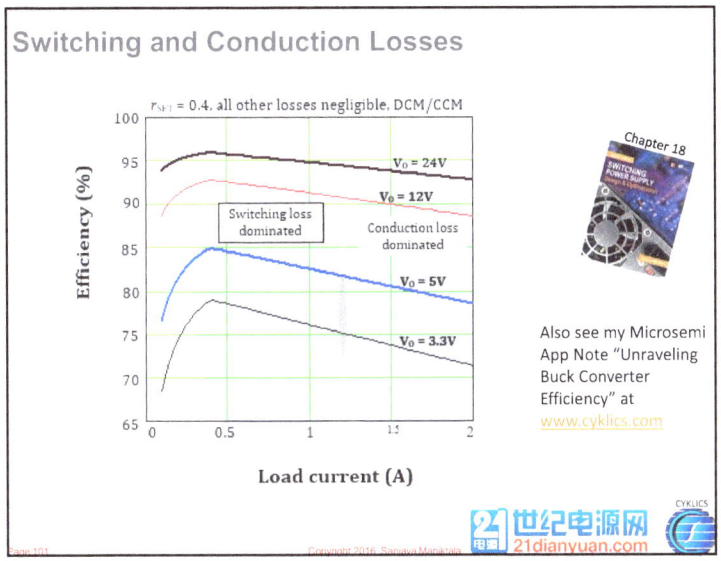

Also see my Microsemi
App Note "Unraveling
Buck Converter
Efficiency" at
www.cyklics.com

101

Overall Efficiency Improvement Direction

- To improve efficiency at max load, lower all **load-dependent** losses such as from Rds of FETs, DCR/ESR, switch transition losses

- To improve efficiency at light loads, lower **load-independent** losses, such as from quiescent current of control IC, driver dissipation. Can also use **pulse-skip mode**!

- Change in topology may help: e.g. **zero-voltage switching (ZVS)**, resonant topologies can reduce switch transition losses ~ 0

102

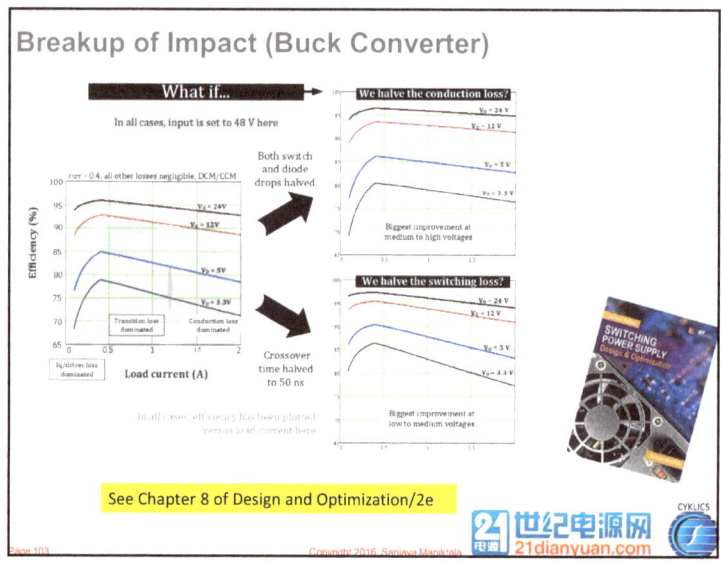

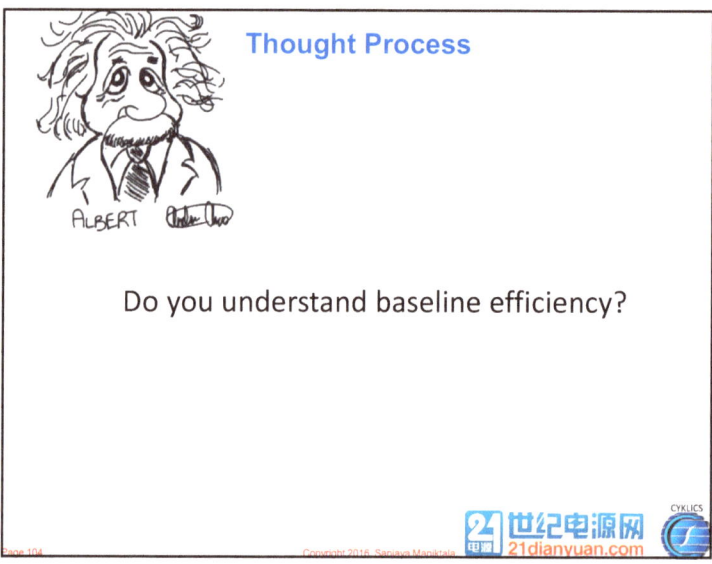

When is Diode Forward Diode Voltage Drop Not a Problem? Rule of Thumb for Efficiency Impact

- The diode drop is <u>not</u> a problem when the output voltage is high! Because the loss in an output diode is Vd*Io. Whereas output wattage (useful power) is Vo*Io. So impact on efficiency by output diodes is of the order of (rule of thumb):

-
$$\Delta\eta \approx \frac{Vd}{Vo}$$

η=Po/Pin (Efficiency)

For example is we use an ultrafast diode with a forward drop of 1.1V on a 24V output, its impact on efficiency is

$$\Delta\eta \approx \frac{Vd}{Vo} = \frac{1.1}{24} = 0.046 \equiv 4.6\%$$

Irrespective of load or input voltage!!!

Whereas, the same diode used for a 5V output would give us:

$$\Delta\eta \approx \frac{Vd}{Vo} = \frac{1.1}{5} = 0.22 \equiv 22\%$$

Irrespective of load or input voltage!!!

105

Best Possible Efficiency!

- The baseline efficiency of any converter is:

$$\eta = \frac{Po}{Pin} = \frac{Vo \times Io}{Vo \times Io + Vd \times Io} = \frac{Vo}{Vo + Vd} \equiv \frac{1}{1 + \frac{Vd}{Vo}}$$

Irrespective of line, load and even topology

This is the best possible efficiency, even if all other losses, such as in the input FET, transformer, capacitors etc is zero. We can't get better than this

Note: the reason we could come up with such a simple formula above is that forward voltage drop across a diode is almost independent of current... think!

106

"Baseline Efficiency"

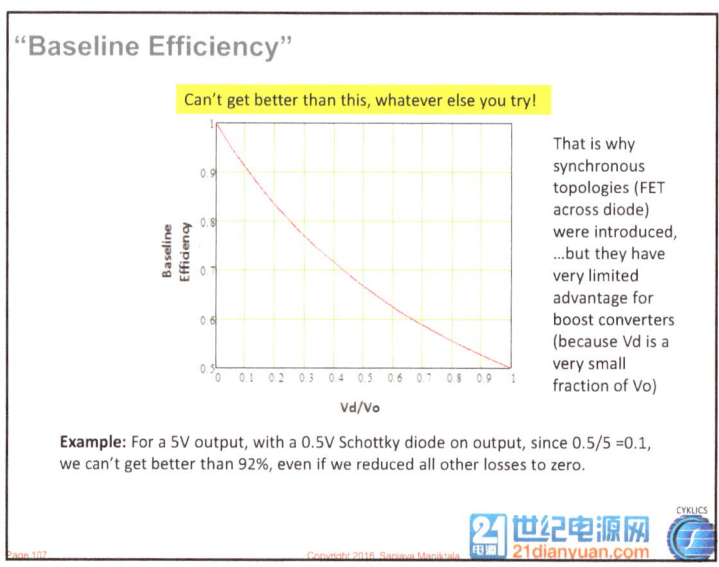

Can't get better than this, whatever else you try!

That is why synchronous topologies (FET across diode) were introduced, ...but they have very limited advantage for boost converters (because Vd is a very small fraction of Vo)

Example: For a 5V output, with a 0.5V Schottky diode on output, since 0.5/5 =0.1, we can't get better than 92%, even if we reduced all other losses to zero.

107

Power Factor Correction "PFC" Front-end (Boost with high output voltages)

The basic purpose of PFC is that the load should appear **as a resistor** to the AC mains...So, as the main voltage rises and falls, so does the current drawn from the mains ...just like a resistor in which since V=IR, so V ∝ I. this ensures power quality.

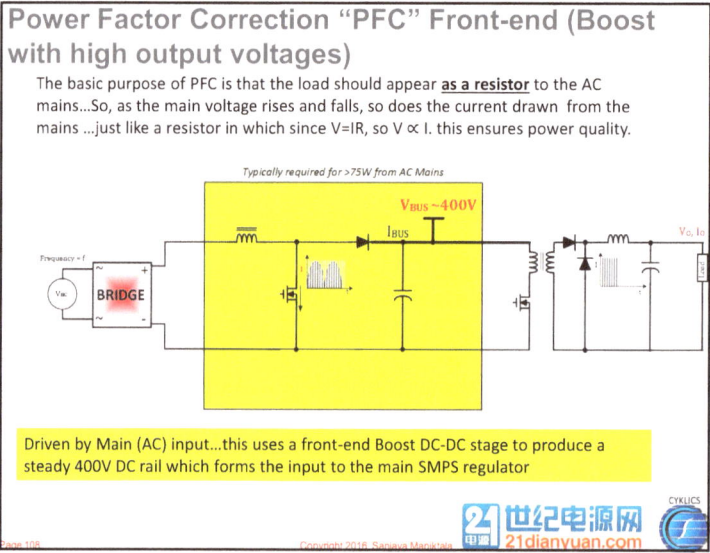

Driven by Main (AC) input...this uses a front-end Boost DC-DC stage to produce a steady 400V DC rail which forms the input to the main SMPS regulator

108

Thought Process

Do you think that if we switch "fast" we will always reduce switching losses (and improve efficiency)?

109

Impact on Efficiency Due to Output (PFC) Diode(s)

- For a "Power Factor Correction (PFC) Stage" where we use a boost stage to output a DC voltage of 400V, the impact on efficiency by a diode of drop 1.1V is 0.3% (irrespective of input)! (1.1/400 = 0.00275) So, **if we use two diodes in series** we get the impact of using two diodes in series as compared to one diode as only 0.3%! Why worry about that?

    *However, in reality, the actual efficiency drop due to one PFC diode versus two such diodes in series in a PFC stage can be as much as 5-10%. In other words, **one** diode is 5-10% **worse** in efficiency **than two diodes in series**!!! Why is that?*

110

Because of Reverse Recovery Time of Diode

- Every diode has a reverse recovery time (in effect). Conventional ultrafast diodes had reverse recovery times of 100-500ns (depending on how quickly we try to turn them OFF).

- If so, it implies that for 100-500ns, the "diode" is not a diode --- it has no reverse blocking capability. This can lead to a huge current spike through the accompanying switch/FET, especially in 400V Boost PFC stages, **leading to a "mysterious" drop in efficiency of ~ 5-10%.**

- In the case of a "synchronous FET" trying to serve as the "diode" of the boost stage, the body diode of the synchronous FET has an even worse reverse recovery time than an ultrafast diode…leading to an even worse drop in efficiency. This leads to several solutions….

111

10-15% DROP IN EFFICIENCY

You cannot see this current spike. Any attempt to measure it using a current probe on the drain with a current loop, reduces the spike….(the measurement is intrusive)

$V_{BUS} \sim 400V$

I_{BUS}

V_o, I_o

Frequency - f

BRIDGE

OFF → ON

Huge spike through FET with voltage still present across it, due to finite reverse recovery of diode, causes ~10-15% loss in efficiency AND CAN DESTROY FET

Also produces huge amount of EMI! Adds to filtering Cost. You might have to **slow down the FET** turn-on to improve efficiency and reduce EMI (put 100 ohms in the Gate!)!

112

Effect of <u>Slowing Down</u> the FET in a Boost PFC

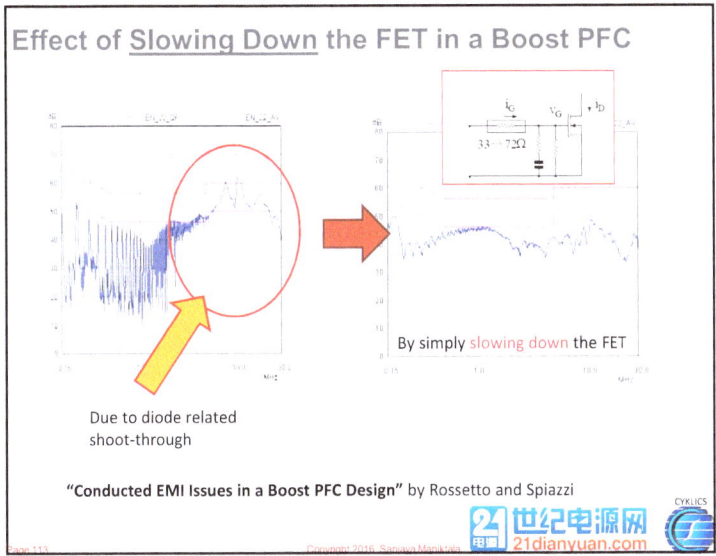

By simply slowing down the FET

Due to diode related shoot-through

"Conducted EMI Issues in a Boost PFC Design" by Rossetto and Spiazzi

113

Ways to correct this: Tandem Diodes (ST Micro)

- Since forward drop of a diode is insignificant for high output voltages, why not use two diode in series?

- The advantage is two 300V diodes in series have much better reverse recovery than a single 600V diode.

- But they should be co-manufactured and co-packaged otherwise they are not dynamically well-balanced (during the short 50-100 ns deadtime, entire reverse voltage may otherwise come across one 300V diode, causing it to fail

114

Ways to correct this: Resonant Snubbers

- Use <u>resonant snubbers</u> which slow the turn-on of the FET and meanwhile, recover the energy from the reverse recovery spike (we <u>do</u> need to extract minority carriers from the diode junction, to make it "recover"). See Fig 20.10 in Design and Optimization 2e.

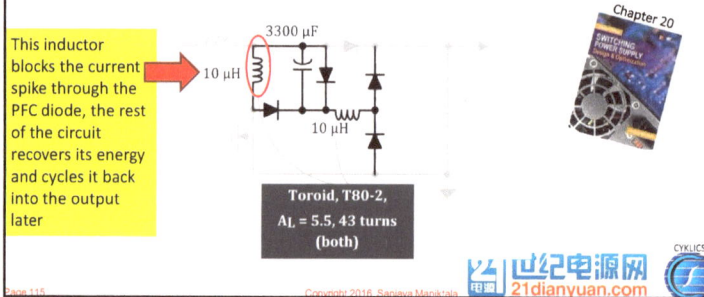

This inductor blocks the current spike through the PFC diode, the rest of the circuit recovers its energy and cycles it back into the output later

115

Ways to correct this: Better PFC diodes

- B) Use Schottky/SiC Diodes? Schottky diodes are "majority carrier" devices which theoretically have "zero reverse recovery" time. But they do have significant body capacitance which in many ways mimics the effect of reverse recovery --- expressed as a "Qrr" or reverse recovery charge.

- C) However, "SiC" diodes do improve efficiency by ~5-10%, without the need for resonant snubbers. Their "Qrr" is a bit suboptimal. But their reverse leakage current (DC) is very low.

- *Keep in mind that Silicon Schottky Diodes (and 600V "Qspeed" Schottky diodes) have a large reverse leakage current which can offset any advantage from better Qrr.*

116

SiC Diodes tout "zero reverse recovery"

Marketing slide from Cree for SiC diodes

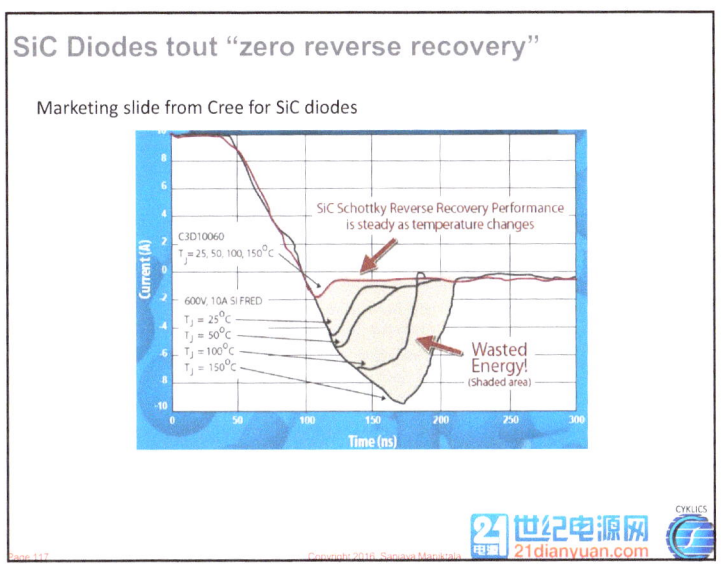

117

Diodes Compared for PFC

• Courtesy NXP

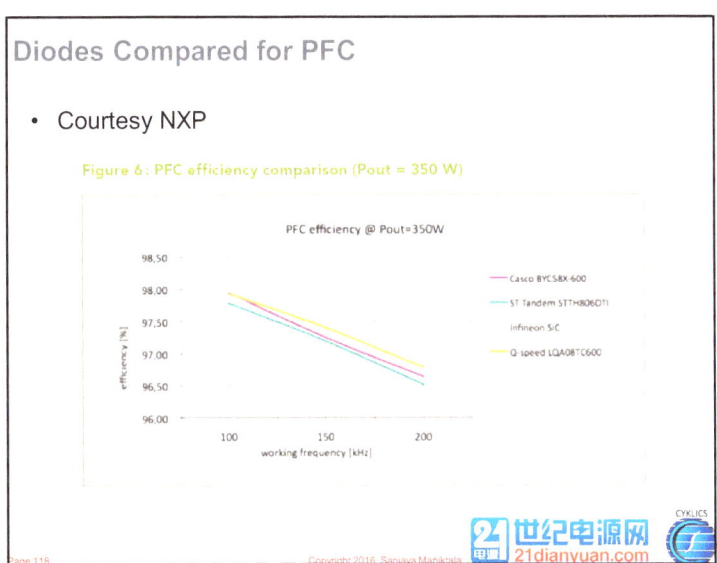

118

119

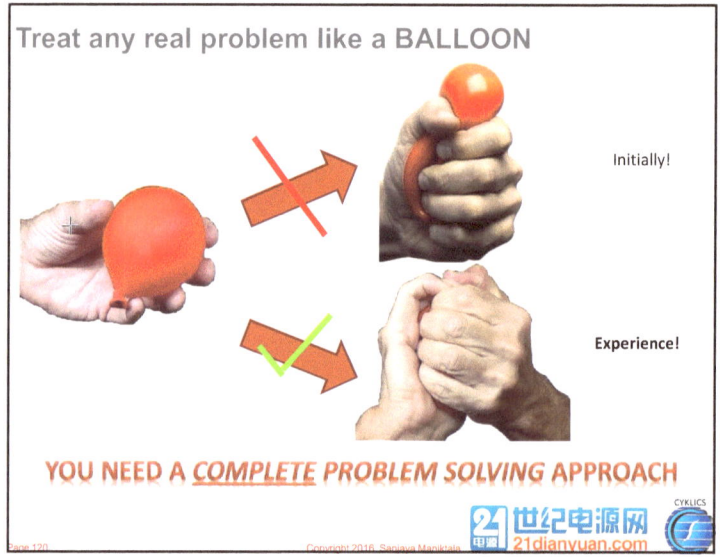

120

Experience helps you look for <u>Side-effects</u>

The truth is: there is almost nothing we can do at a certain point without affecting something else, *somewhere*!

Experience (yours, mine and others') will tell you ***where exactly*** that "side-effect" occurs.

So you can fix it proactively **in the lab**....

...before **your poor customer discovers them for you!**.....

121

To become a successful <u>Power</u> Designer…

Remember: Power supply design is all about *tradeoffs. It involves carefully considered compromises and **optimization** of cost, performance, reliability…*

122

Think of yourself as a master _juggler_

UL: American safety test agency
CE: European Conformity Mark
MTBF: Mean time between failures
DFM: Design for manufacturability

123

Bay Area~1998

124

Failed Assumption during "Worst-case" Testing

http://www.planetanalog.com/document.asp?doc_id=527343

- It was **assumed** that the way to do a "worst-case test" was to load the PC power supply to maximum power on all outputs, raise the temperature and monitor any failures.

- But after passing *all internal tests*, failures occurred in large numbers ….*in Japan*! Because they were in the habit of putting their computers in "standby mode" in the evening --- in which case the main 5V and 12V output rails of the power supply were OFF, and only the *5W standby power supply remained alive*. But since the fans were run off the 12V rail, now the TOPSWITCH was getting too hot as it had no fan and no heatsink (it had seemed to designers that no heatsink was required as it wasn't, during max load testing!)
  But as an integrated switcher, there should have been thermal protection built in!

Page 125   Copyright 2016, Sanjaya Maniktala

125

Pay attention to datasheets: What they are NOT saying or telling you is equally important

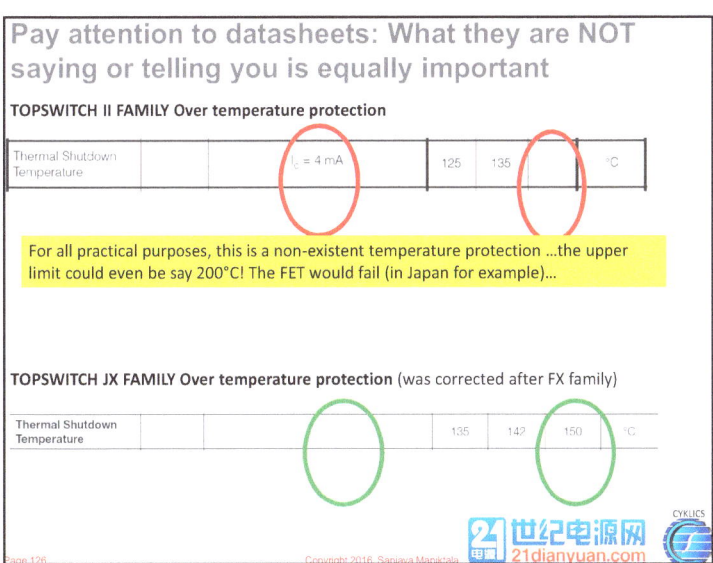

TOPSWITCH II FAMILY Over temperature protection

Thermal Shutdown Temperature		$I_C = 4\ mA$	125	135		°C

For all practical purposes, this is a non-existent temperature protection …the upper limit could even be say 200°C! The FET would fail (in Japan for example)…

TOPSWITCH JX FAMILY Over temperature protection (was corrected after FX family)

Thermal Shutdown Temperature			135	142	150	°C

Page 126   Copyright 2016, Sanjaya Maniktala

126

Pay Attention to *User Habits* and Market Preferences

- Japanese customers had another wishlist....I was told *"they don't mind a power supply failing occasionally...but they do not want failure to disturb them with loud noises in the work area."*

- So we had to put <u>small fuses in the drain side of every FET</u>, just to ensure that under a fault condition, the fuses failed softly (open)...before the FET itself erupted with a loud noise! Adds cost, but it is necessary in this case!

127

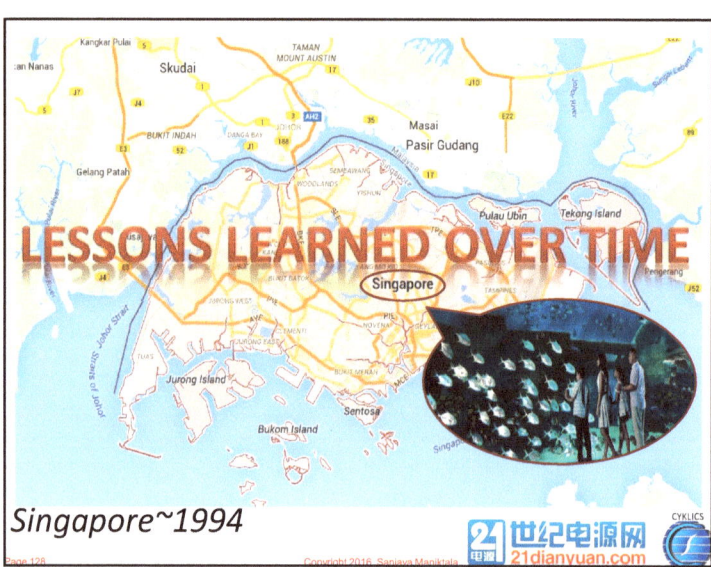

Singapore~1994

128

Multiple Concerns Addressed Successfully

- This was a 65W-70W power supply for either Apple or HP.

- It reflects a keen problem-solving approach in the face of an unexpected setback, which eventually balanced multiple aspects such as:

a) *Safety/Certification UL/CE*
b) *Reliability (MTBF)*
c) *Cost ($, Yuan)*
d) *Performance (e.g. Efficiency)*

Copyright 2016, Sanjaya Maniktala

129

UL Test, Step 1 (FET fails shorted, in Standard UL Abnormal Testing)

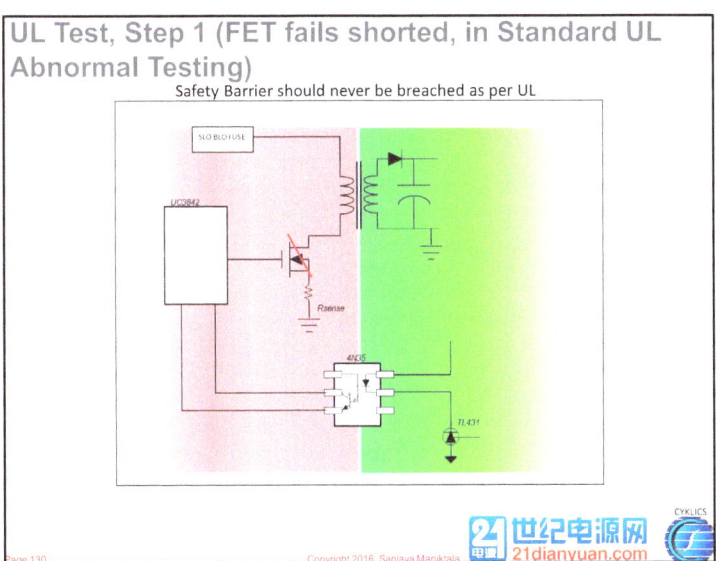

Safety Barrier should never be breached as per UL

Copyright 2016, Sanjaya Maniktala

130

UL Test, Step 2 (Current in Inductance pushes through Sense Resistor)

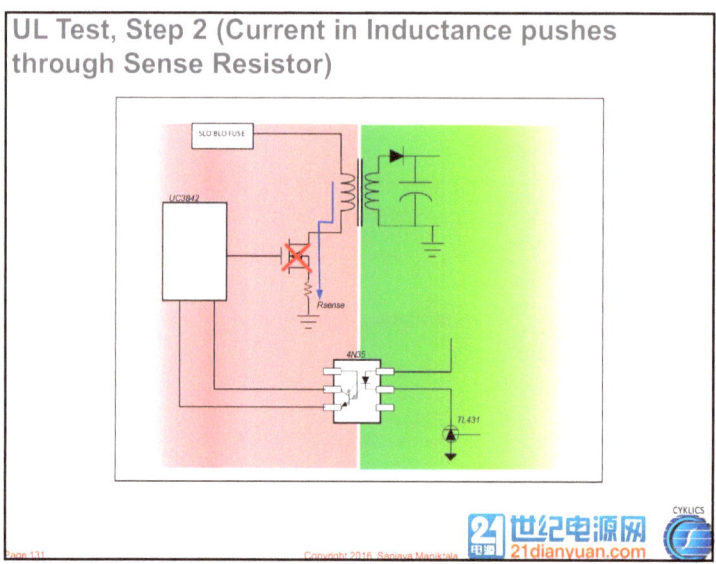

131

UL Test, Step 3 (Rsense fails open, current diverts into Control IC)

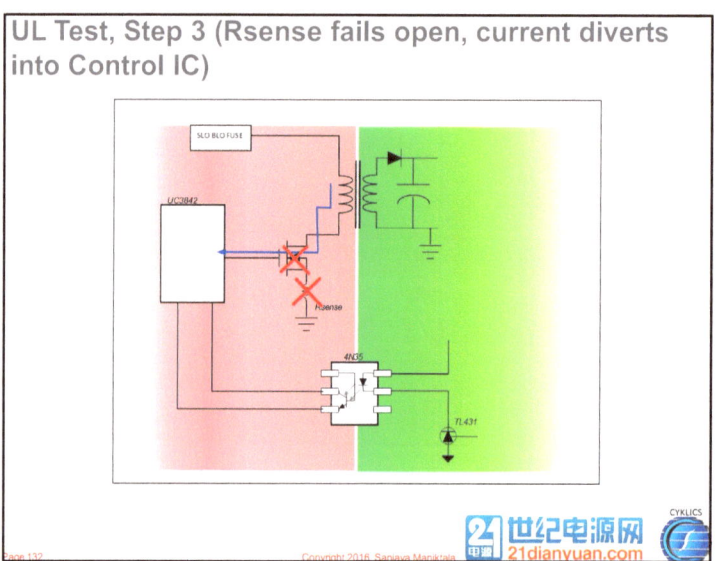

132

UL Test, Step 4 (Control IC fails short)

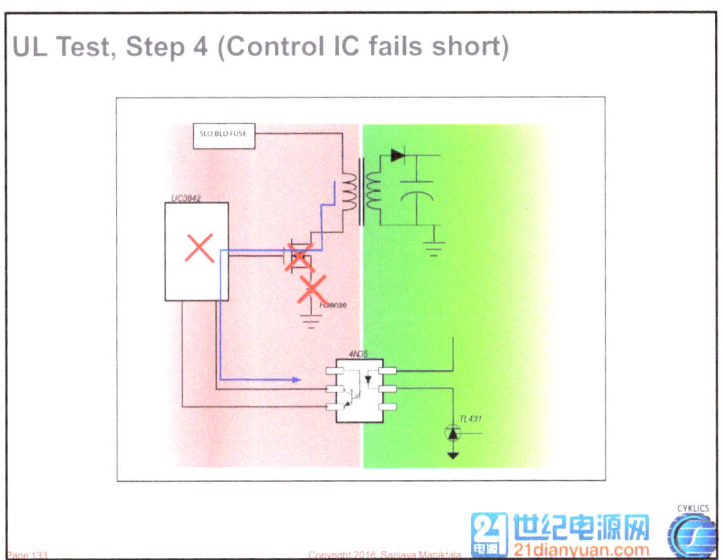

133

UL Test, Step 5 (Optocoupler fails short, then erupts)

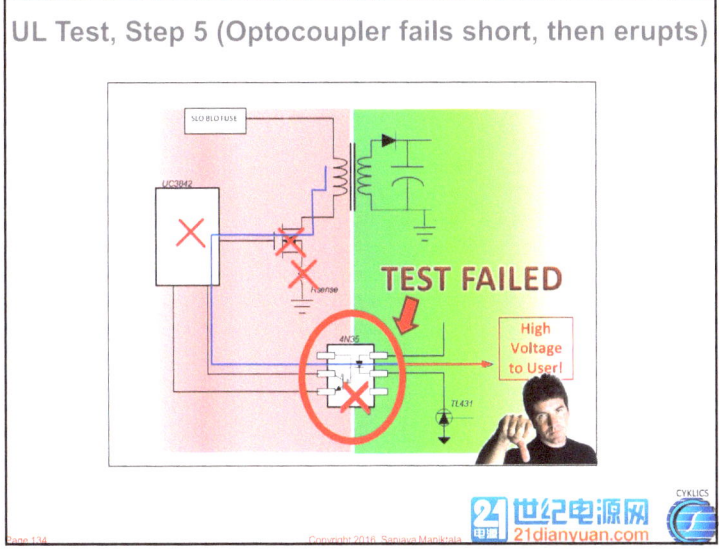

134

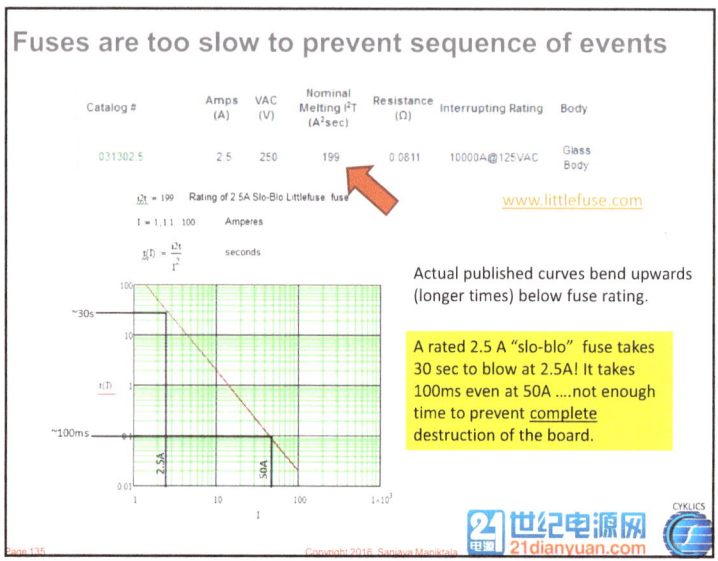

135

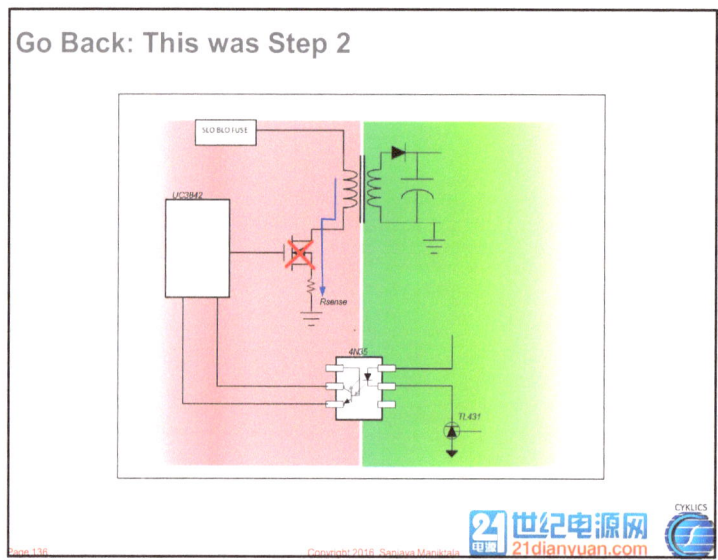

136

New UL Test, Step 3 (Rsense fails open, current diverts into Gate-Source Zener)

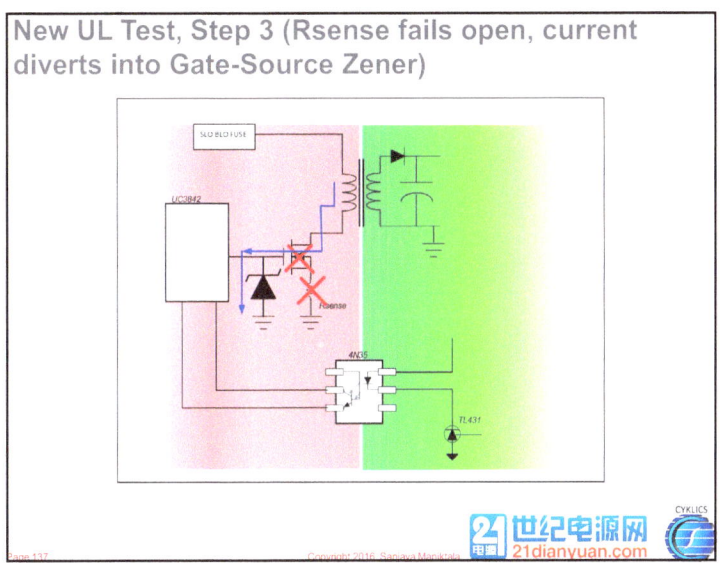

137

Nrw UL Test, Step 4 (Zener **fails short**, current continues through zener), not in IC

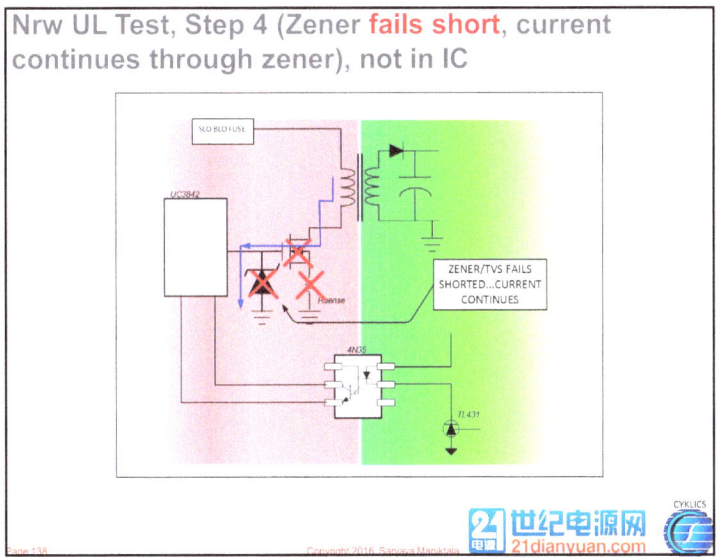

138

New UL Test, Step 5 (Input Fuse blows, no violation of integrity of safety barrier occurred)

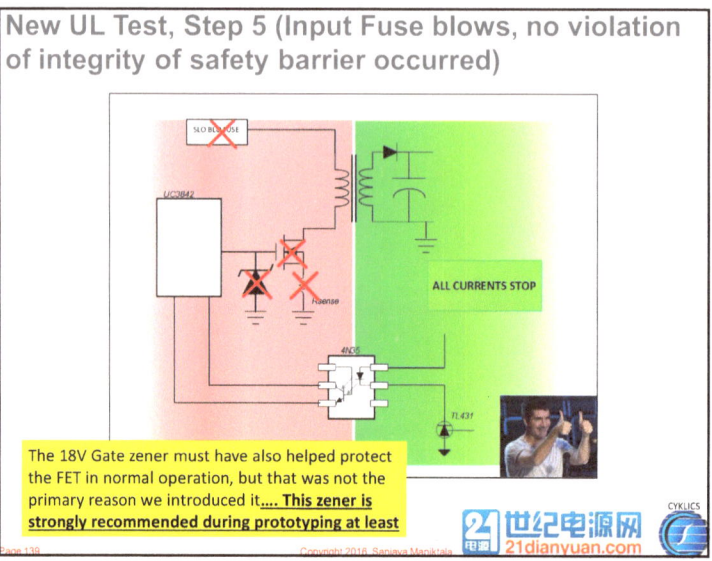

The 18V Gate zener must have also helped protect the FET in normal operation, but that was not the primary reason we introduced it**.... This zener is strongly recommended during prototyping at least**

139

Compaq designs we had built in Bombay (Mumbai)

- Schematics from **COMPAQ** always had a "mysterious TVS" across the sense resistor! Now I knew why. But a 500mW Gate-Source zener was much cheaper than a TVS.

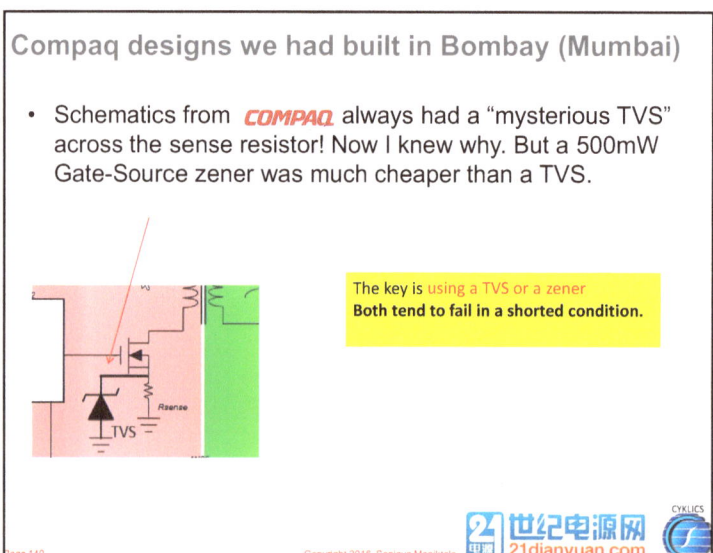

The key is using a TVS or a zener
Both tend to fail in a shorted condition.

140

Thought Process

Do you think all 5A/60V Schottky diodes
are the same? Have you heard about
dV/dt rating?

141

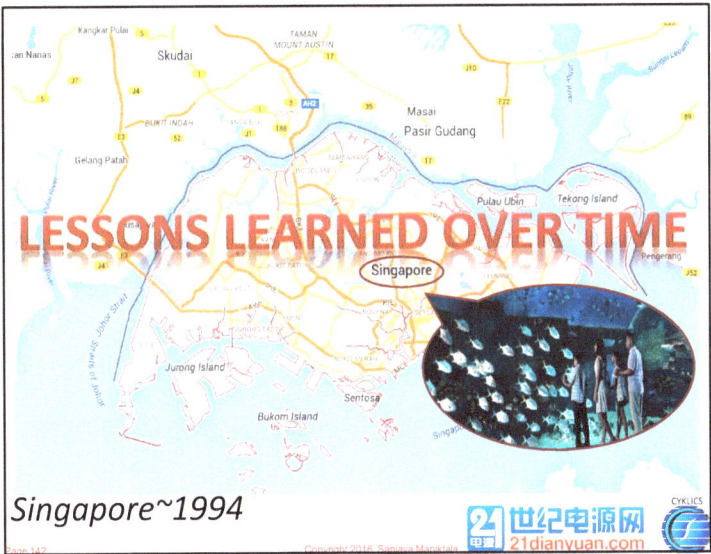

Singapore~1994

142

Schottky dV/dt failures (due to "bad" layout)

One or two Schottky diodes in several thousand were failing "mysteriously" in pre-production testing. That was a huge and unacceptable "ppm" (parts per million) failure rate

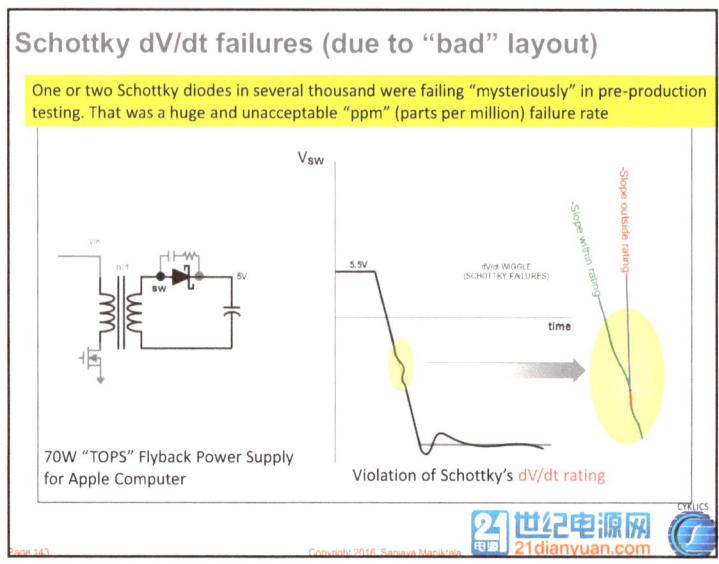

70W "TOPS" Flyback Power Supply for Apple Computer

Violation of Schottky's dV/dt rating

143

A Ferrite bead to the rescue

The Schottky diode had been pulled away to the side for mounting on chassis for thermal reasons. This was causing the ringing. Finally, a lossy (Ni-Zn) ferrite bead was mounted to "kill" the ringing..

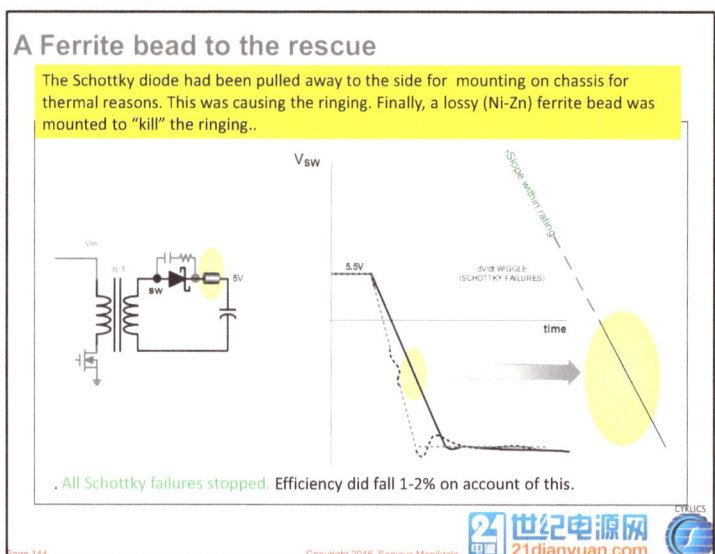

. All Schottky failures stopped. Efficiency did fall 1-2% on account of this.

144

Schottky diodes have a crucial dV/dt rating!

- Most modern Schottky diodes can withstand dV/dt of 10,000V/us. But this was a cheaper "equivalent" diode from ST Micro, with a rating of only 2000V/us. The other part was from IR and was exhibiting no failures.

- **DO NOT FORGET TO ASK THE VENDOR FOR THE "dV/dt RATING OF HIS SCHOTTKY"!**

145

ALBERT

Thought Process

Do you think all ceramic capacitors are about the same?

146

Tolerances, Drifts, Aging of Ceramic Caps

TCC (Thermal coefficient of capacitance) :

Low temperature limit of range(°C)	Upper temperature limit of range (°C)	Maximum allowable change in capacitance from 25°C (at 0 VDC, over entire operating temperature range)
X = -55	4 = 65	F = ± 7.5 %
Y = -30	5 = 85	P = ±10 %
Z = -10	6 = 105	R = ± 15 %
	7 = 125	S = ± 22%
	8 = 150	T = +22, -33 %
		U = +22, -56 %
		V = +22, -82 %

Check difference between X5R, X5S, Z5U, Y5V etc

Page 147

147

Actual Capacitance of a "real" ceramic cap

- Capacitance as specified in a datasheet is usually measured at an applied voltage of 1 VRMS, at 1 kHz and 25 °C. In an actual circuit we could see the following worst-case spread (beyond 100 kHz, with the possibility of both AC and DC voltages being considered):

- *COG.* Initial tolerance (±5%), TCC (±0.15%), voltage stability (0%), frequency stability (0%), aging (0%). Combining, we get worst-case upper limit as C×1.05×1.0015=C×1.0516, i.e. 5.16% higher. Similarly, the worst-case lower limit is calculated by C×0.95×0.9985=C×0.94858, i.e. 1-0.94858=0.0514, or 5.14% lower. The calculation is not so obvious, but truly reflects the worst-case *combinational* drift. Finally, C → *+5.16, -5.14%*. Note that a small non-zero TCC is already included within its initial tolerance range.

- *X7R.* Initial tolerance (±10%), TCC (+2, -10%), voltage stability (+15, -10%), frequency stability (+5, -15%), aging (-3%). Combining, we get worst case C → *+35%, -40%.*

- *Z5U.* Initial tolerance (±20%), TCC (+2, -54%), voltage stability (+22, -56%), frequency stability (+5, -15%), aging (-25%). Combining, we get C → *+57%, -90%.*

And most of this is not even guaranteed: just by typical curves!

Page 148

148

Caution when using capacitors

Ceramic capacitors: may need to use twice the capacitance "calculated". Can cause immediate destruction of low voltage switchers (next slide)

Aluminum electrolytic capacitors: Polarity. Life halves every 10°C rise in temperature. Careful life prediction necessary and also PCB layout (keep away from hot components). ESR climbs steeply below 25°C, unusable. Higher leakage.

Film capacitors: More stable than ceramic. If used, prefer Polpropylene for pulsed applications (with high dV/dt), such as for snubbers and clamps, not Polyester (Mylar) for example

Tantalum capacitors: Polarity. Because of surge current heat buildup, may need to use to half rated voltage usually …. Can erupt badly.

Solid Polymer capacitors: Polarity. Very low ESR, stable. Up to 25V. Tend to fail shorted.

Hybrid (Solid/Liquid) Polymers capacitors: Very low ESR, stable. Up to 100V. Tend to fail open.

149

STOP: Reality Check

HOW GOOD ARE YOUR LAB SKILLS?

150

The Biggest Lie of Electrical Schematics

- A ground is shown everywhere with the same symbol. In reality the grounds can be very different due to resistive drops where high currents are flowing.
- Nowadays chips try to present separate "analog" and "power" ground pins. In reality we usually connect them together very close to the IC on a large copper plane.
- A ground loop can induce huge circulating currents, quite like the paralleled windings case. Here the driving force is the voltage gradient along the ground by two different paths, forcing the "excess" voltage to be dropped across the very small PCB trace resistances.

But what about inadvertent ground loops?

151

Bay Area~2004

152

How "ground loops" can show up

Around 2004 (at Nat Semi): I noticed extremely high output ripple and traced the cause to this..

The second probe wasn't even "connected"!

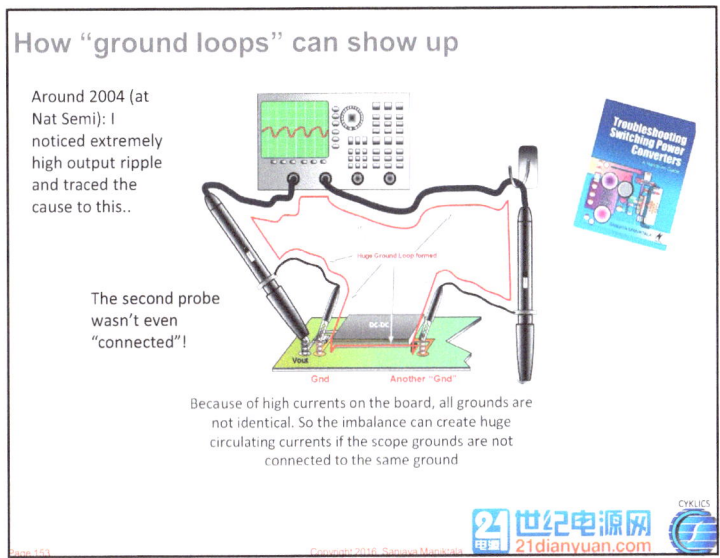

Because of high currents on the board, all grounds are not identical. So the imbalance can create huge circulating currents if the scope grounds are not connected to the same ground

153

Lingering Issues with Output Noise and Ripple Measurement

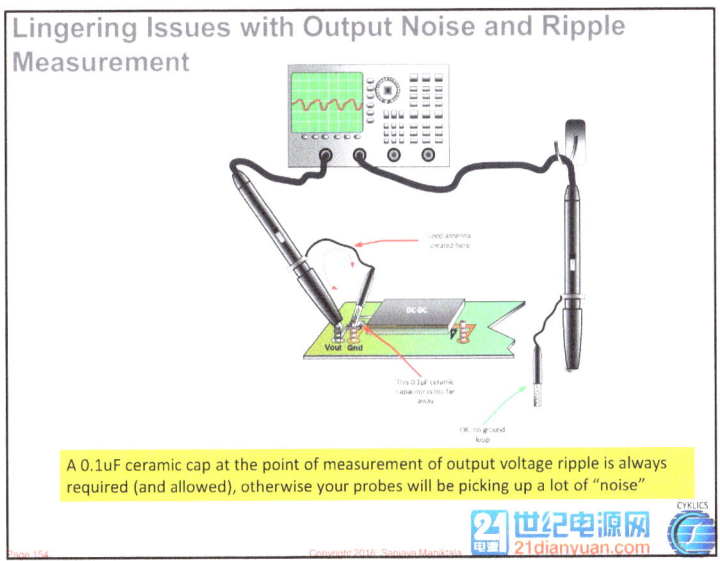

A 0.1uF ceramic cap at the point of measurement of output voltage ripple is always required (and allowed), otherwise your probes will be picking up a lot of "noise"

154

Avoid PCB Ground Loops too (that's obvious)

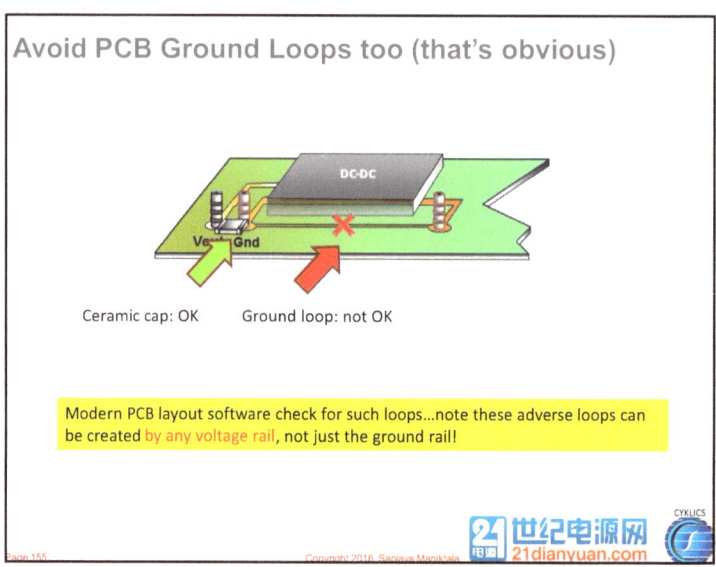

Ceramic cap: OK   Ground loop: not OK

Modern PCB layout software check for such loops...note these adverse loops can be created by any voltage rail, not just the ground rail!

155

Customer reporting "Very High Output Ripple": CHECK

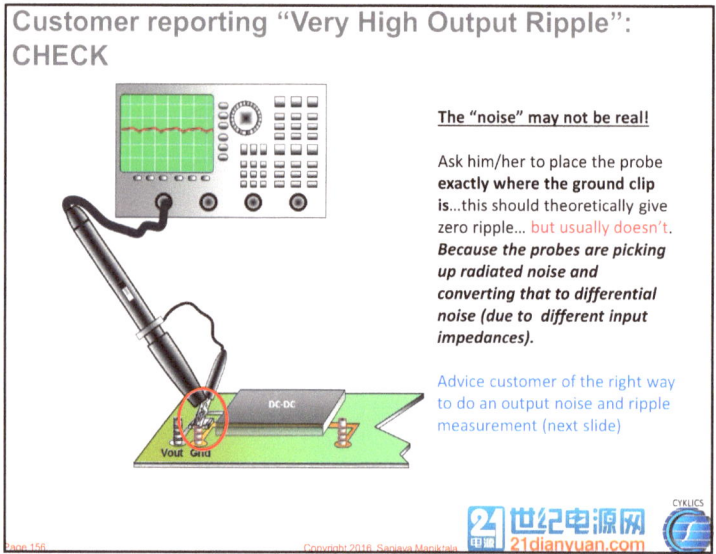

The "noise" may not be real!

Ask him/her to place the probe **exactly where the ground clip is**...this should theoretically give zero ripple... but usually doesn't. *Because the probes are picking up radiated noise and converting that to differential noise (due to different input impedances).*

Advice customer of the right way to do an output noise and ripple measurement (next slide)

156

Correct Output Noise and Ripple Measurement

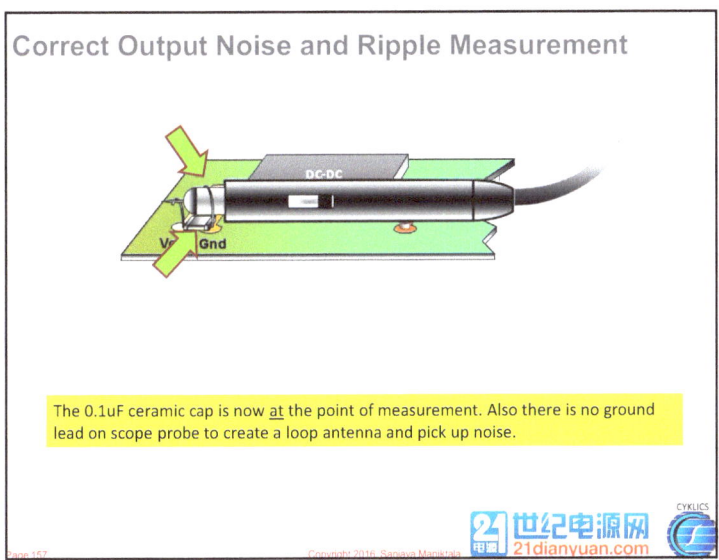

The 0.1uF ceramic cap is now <u>at</u> the point of measurement. Also there is no ground lead on scope probe to create a loop antenna and pick up noise.

157

"Load card" must also have a 100nF for output Noise and Ripple testing (maybe disc ceramic)

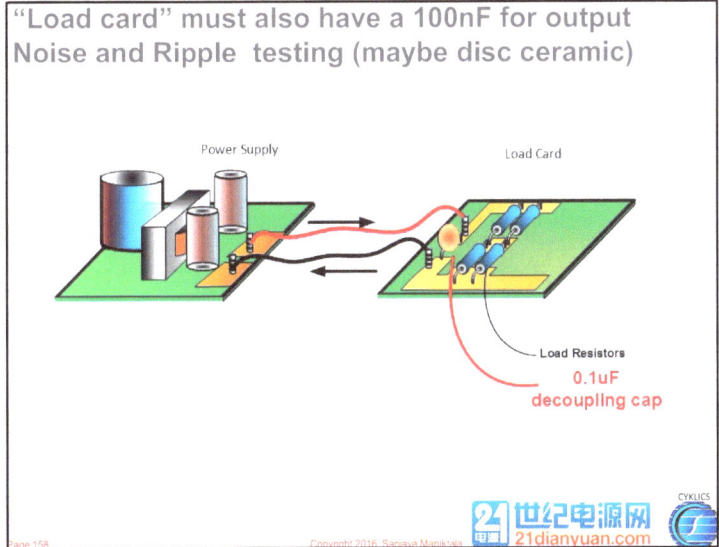

158

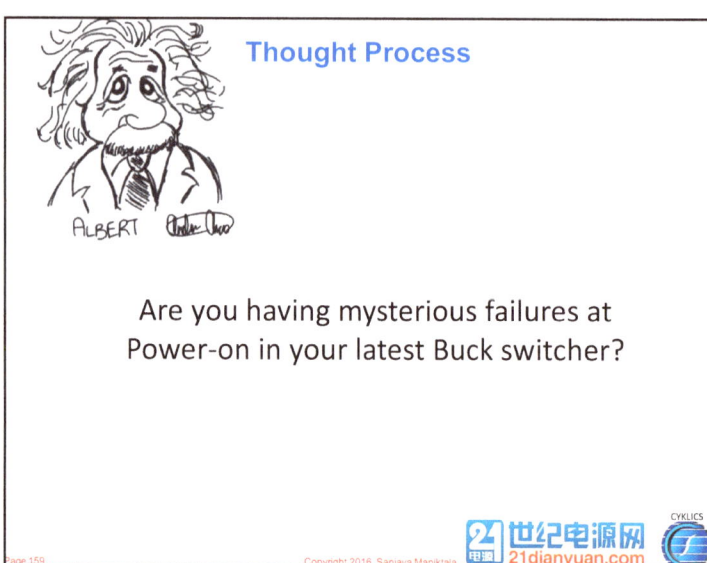

159

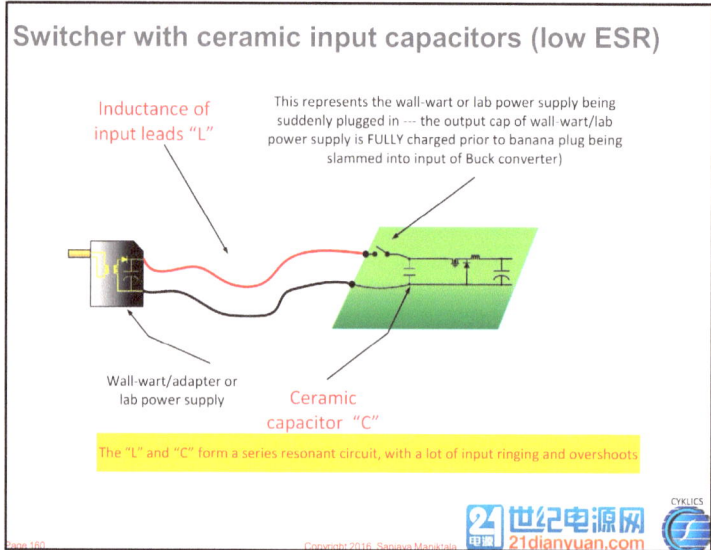

160

Voltage at input of switcher with ceramic input cap

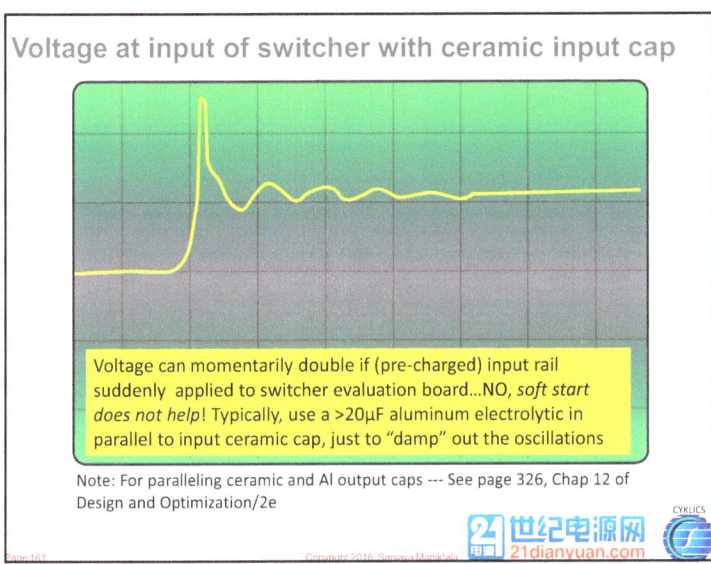

Voltage can momentarily double if (pre-charged) input rail suddenly applied to switcher evaluation board...NO, *soft start does not help*! Typically, use a >20μF aluminum electrolytic in parallel to input ceramic cap, just to "damp" out the oscillations

Note: For paralleling ceramic and Al output caps --- See page 326, Chap 12 of Design and Optimization/2e

161

How is the difference between a Forward and Flyback transformer implemented?

- By the relative polarity of the windings!

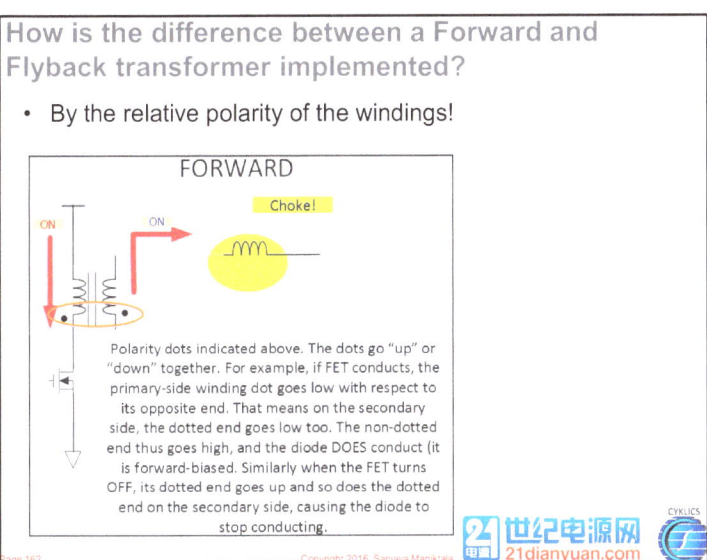

FORWARD

Choke!

ON

ON

Polarity dots indicated above. The dots go "up" or "down" together. For example, if FET conducts, the primary-side winding dot goes low with respect to its opposite end. That means on the secondary side, the dotted end goes low too. The non-dotted end thus goes high, and the diode DOES conduct (it is forward-biased. Similarly when the FET turns OFF, its dotted end goes up and so does the dotted end on the secondary side, causing the diode to stop conducting.

162

Basic debugging of new transformer/prototype

- **Check the polarity of the windings!** Too often, a new Flyback/Forward transformer comes with a wrong polarity and you spend days on it before you realize it. If the polarities are correct, this is what should happen by a simple experiment (on any transformer):

- Connect the main primary winding to the main secondary winding with the presumed non-dotted end of the secondary, connected to the dotted end of the primary. The inductance L2 should be measured to be at least slightly greater than L1 if things are right! (see next slide)

163

Basic debugging of new transformer/prototype

Connection

The inductance L2 should be measured to be at least slightly greater than L1 if things are right!

Don't forget to check any "auxiliary (IC power) winding" too.

164

Leipzig~1995-97

165

The 3-phase 7.5kW "Rectifier" project (1995-1997)

- For 2 years, a team of brilliant German engineers beside me, worked on a high-power 3-phase project, which essentially combined (OR-ed) three 1-phase Power Factor Correction stages:

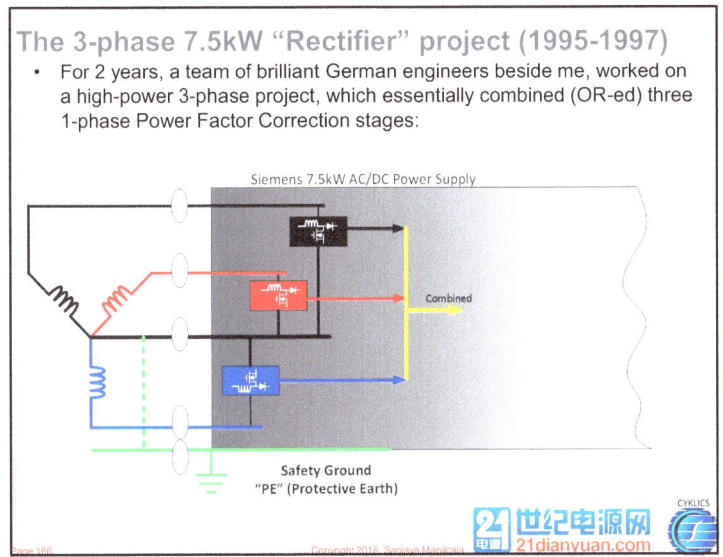

166

This was intended/required/assumed

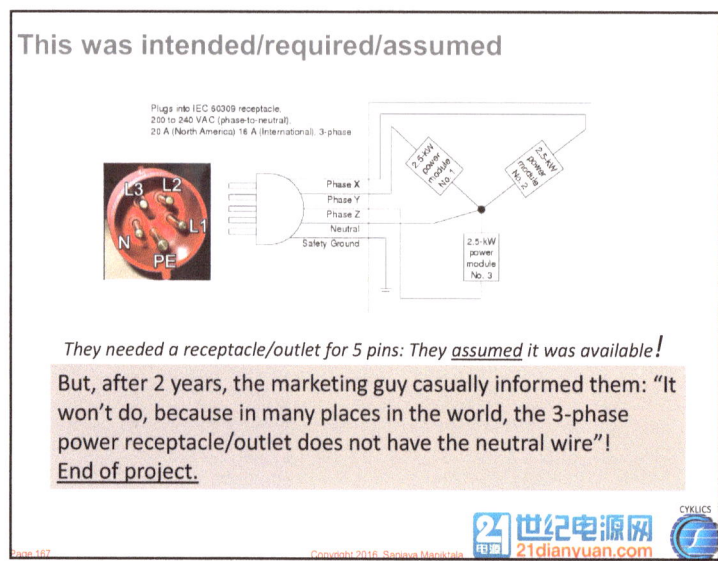

Plugs into IEC 60309 receptacle.
200 to 240 VAC (phase-to-neutral).
20 A (North America) 16 A (International). 3-phase

They needed a receptacle/outlet for 5 pins: They <u>assumed</u> it was available!

But, after 2 years, the marketing guy casually informed them: "It won't do, because in many places in the world, the 3-phase power receptacle/outlet does not have the neutral wire"!
<u>End of project.</u>

167

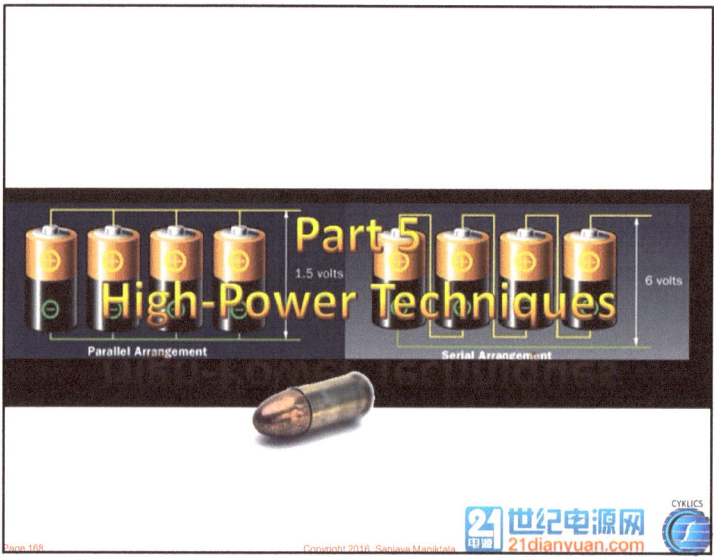

168

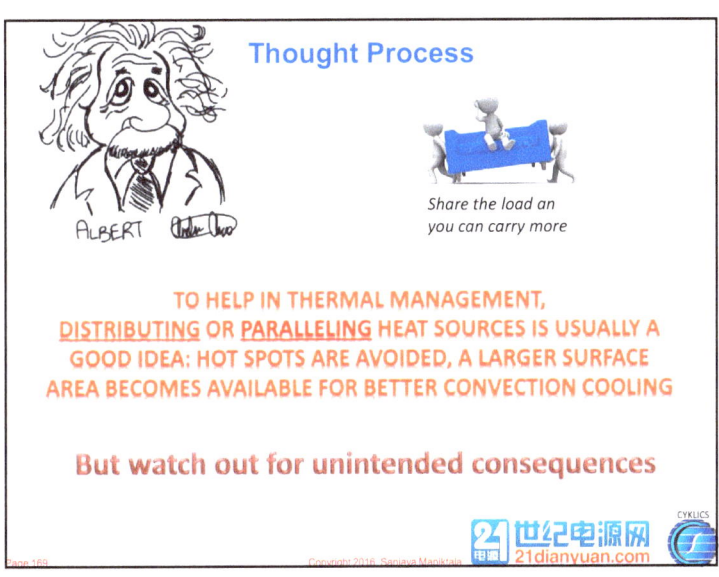

169

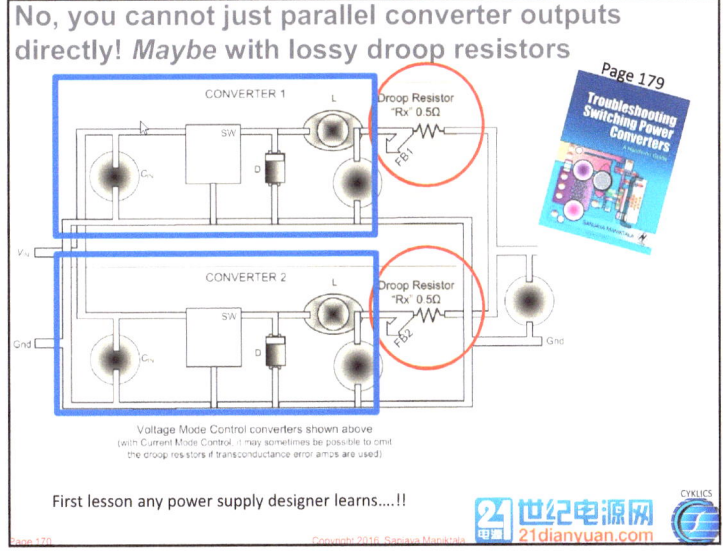

170

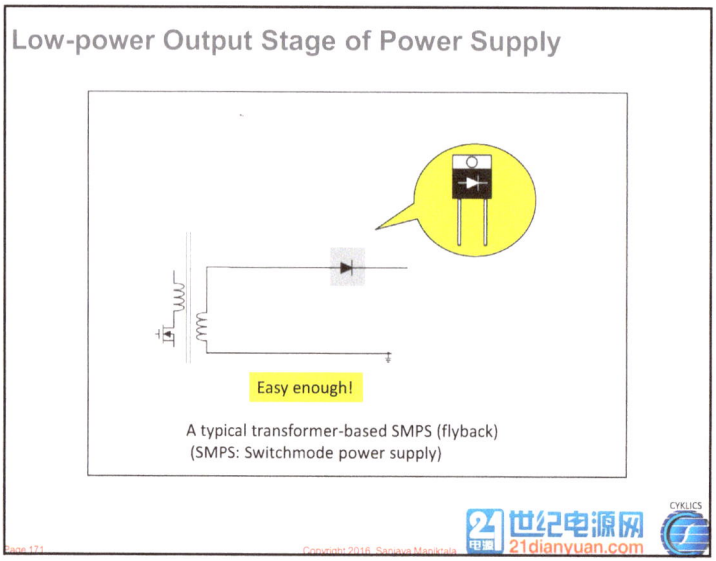

171

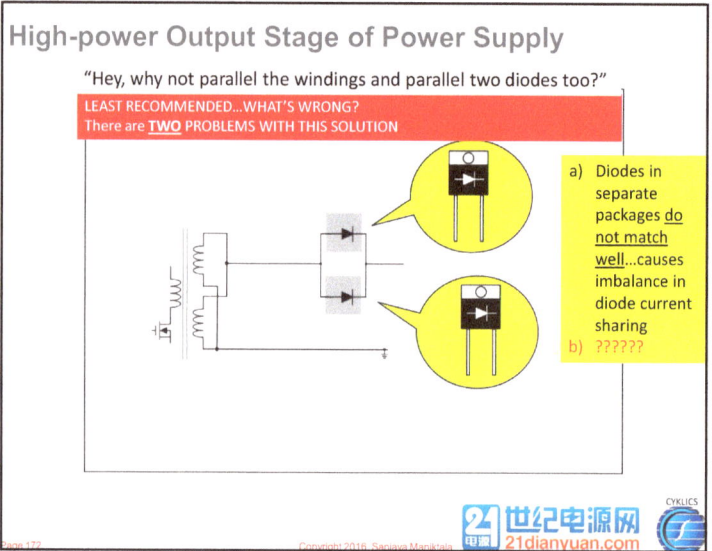

172

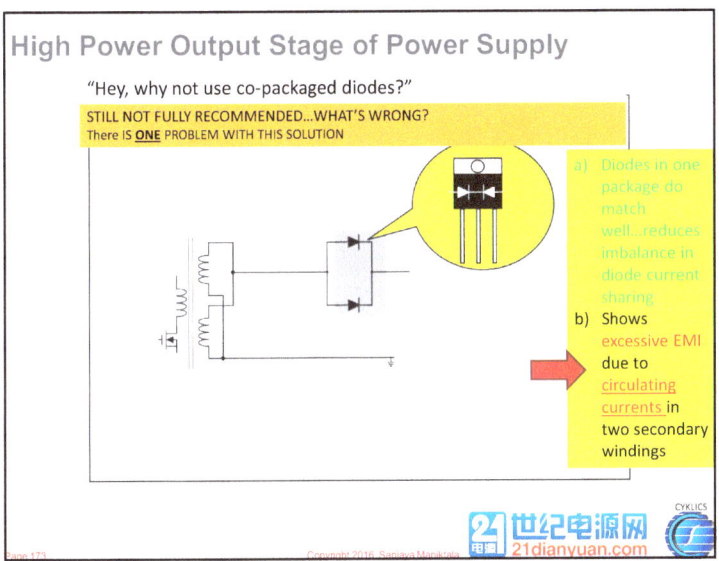

High Power Output Stage of Power Supply

"Hey, why not use co-packaged diodes?"

STILL NOT FULLY RECOMMENDED...WHAT'S WRONG?
There IS **ONE** PROBLEM WITH THIS SOLUTION

a) Diodes in one package do match well...reduces imbalance in diode current sharing

b) Shows excessive EMI due to circulating currents in two secondary windings

173

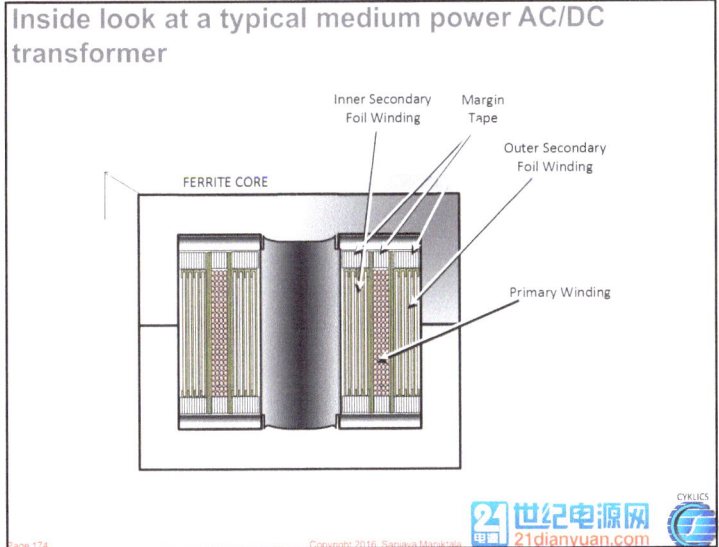

Inside look at a typical medium power AC/DC transformer

FERRITE CORE

Inner Secondary Foil Winding

Margin Tape

Outer Secondary Foil Winding

Primary Winding

174

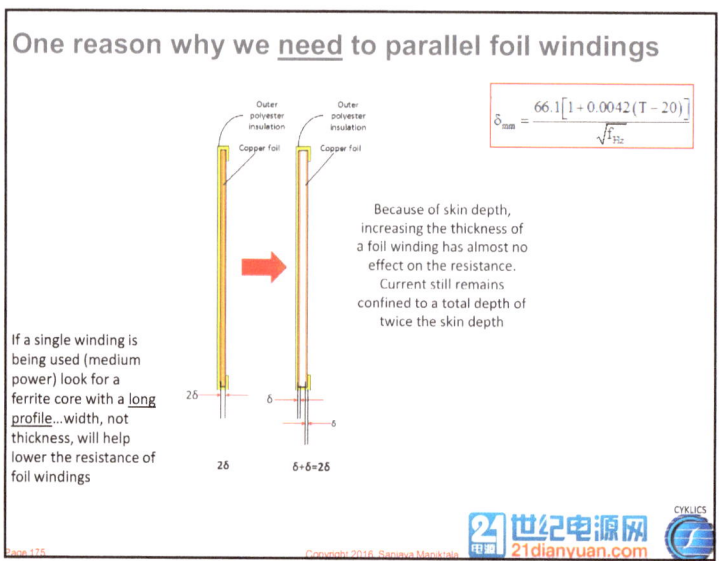

175

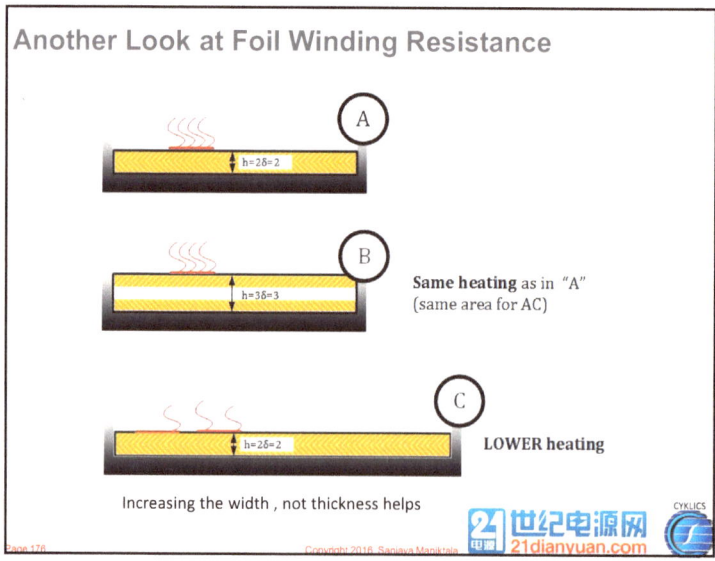

176

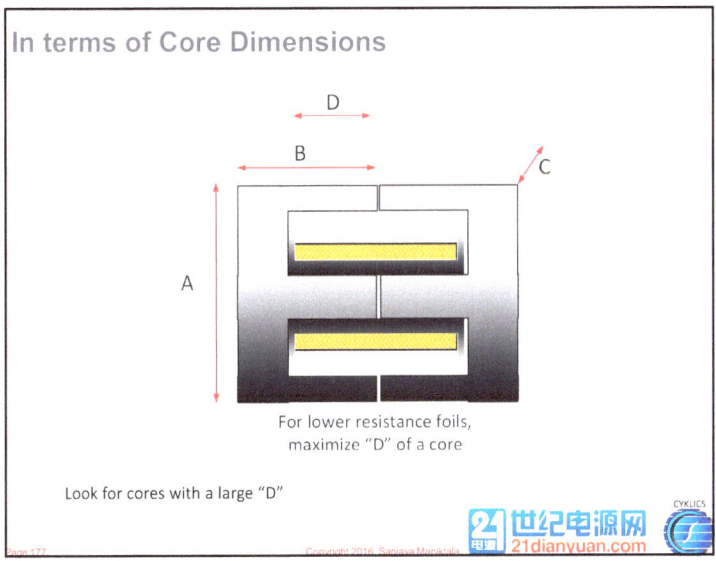

177

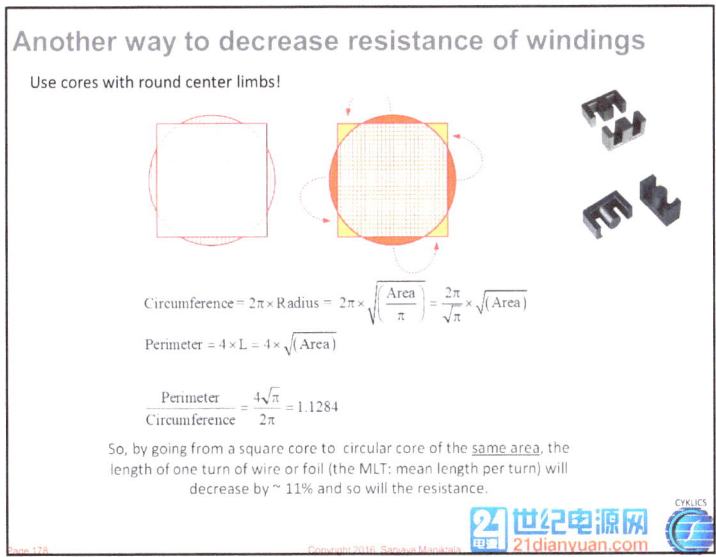

178

Sandwiched Windings

- Sandwiched windings reduce *leakage inductance* and also reduce proximity (AC) losses in the copper.

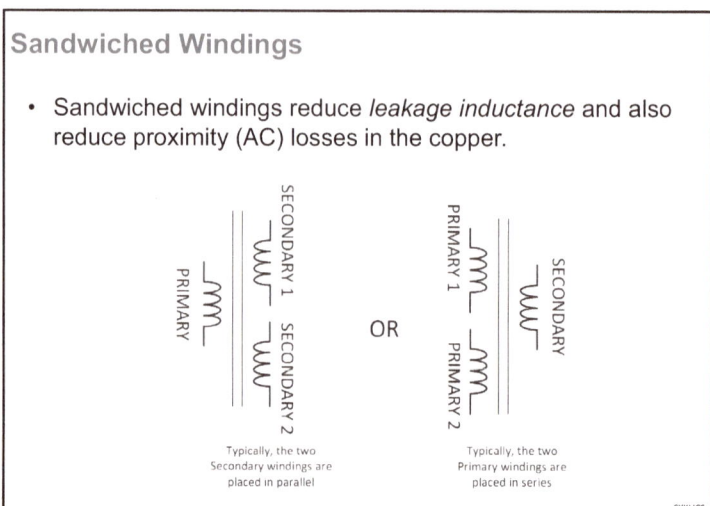

The problem with directly paralleling windings

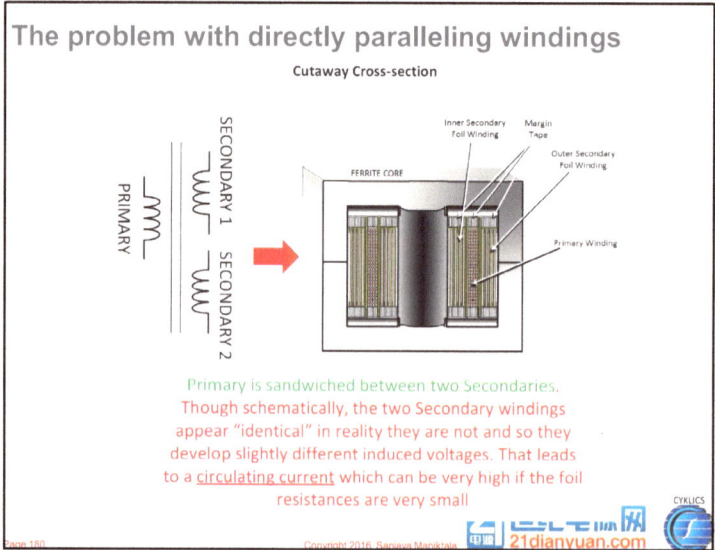

Primary is sandwiched between two Secondaries. Though schematically, the two Secondary windings appear "identical" in reality they are not and so they develop slightly different induced voltages. That leads to a circulating current which can be very high if the foil resistances are very small

(Real) Paralleled Windings vs. *Simulations*

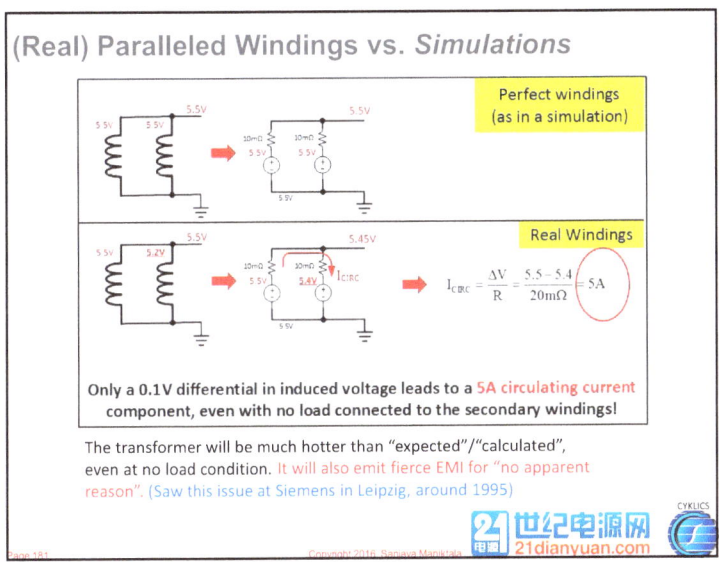

Only a 0.1V differential in induced voltage leads to a **5A circulating current** component, even with no load connected to the secondary windings!

The transformer will be much hotter than "expected"/"calculated", even at no load condition. It will also emit fierce EMI for "no apparent reason". (Saw this issue at Siemens in Leipzig, around 1995)

181

High-power Output Stage of Power Supply

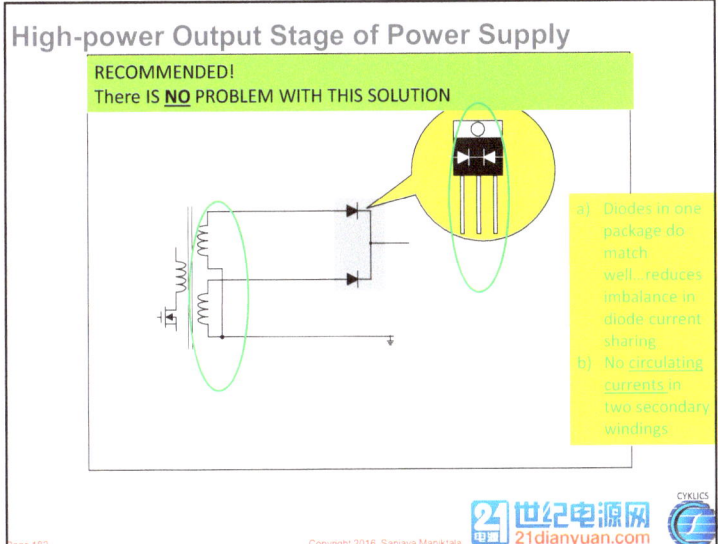

RECOMMENDED!
There IS **NO** PROBLEM WITH THIS SOLUTION

a) Diodes in one package do match well.. reduces imbalance in diode current sharing
b) No circulating currents in two secondary windings

182

Lessons of co-packaged devices

- Components produced in the same batch, preferably on the same wafer (co-packaged + co-manufactured semiconductors) have very good <u>relative</u> tolerances and matched temperature coefficients too (besides the same ambient temperature too, in a real application).

- In the same package, though the thermal management is worse than with separate components, both co-packaged devices see the same temperature, so their *relative drift* due to "tempco" is very low.

STOP: Check your progress

PARALLELING BRIDGE RECTIFIERS

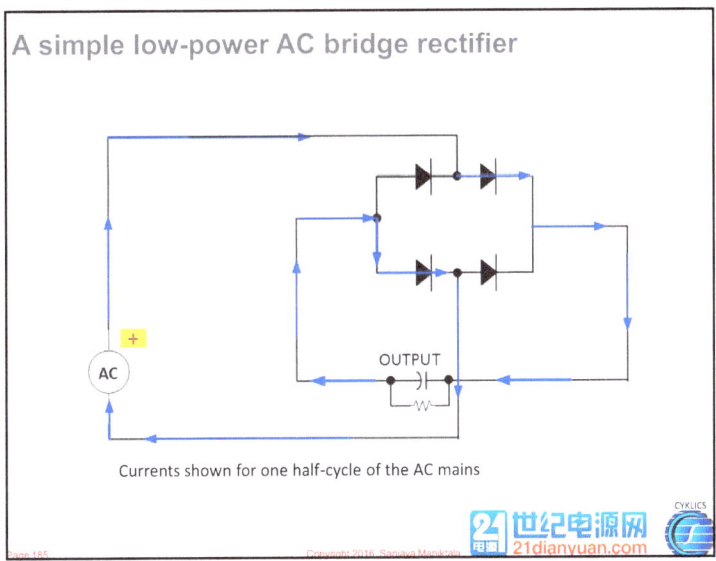

185

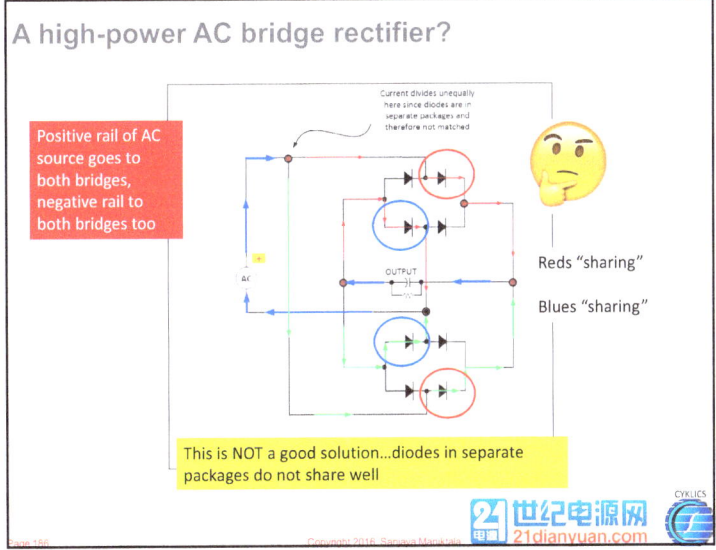

186

A high-power AC bridge rectifier (<u>best</u> solution)

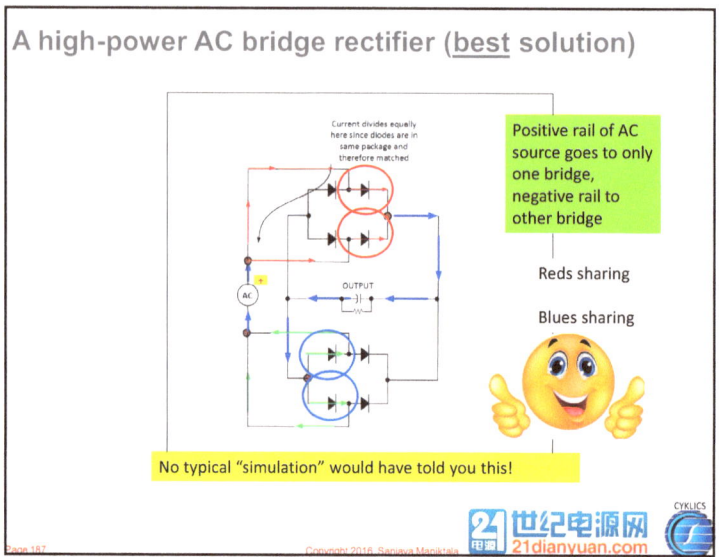

Current divides equally here since diodes are in same package and therefore matched

Positive rail of AC source goes to only one bridge, negative rail to other bridge

Reds sharing

Blues sharing

No typical "simulation" would have told you this!

187

The Temperature Coefficients (tempco) of other devices

- Bipolar transistors' Vce(sat) have negative temperature coefficient for low collector currents, but positive tempco at high currents. Hard to parallel effectively.

- Rds of FETs has a positive tempco. Easy to parallel.

- Low-voltage zeners have a *negative* temperature coefficient. High-voltage zeners have a *positive* tempco. Somewhere in the middle, around 5.1 to 5.6V, the tempco flips sign and passes through zero. That is why it is often said that a 5.1 or 5.6 Volt discrete zener is a good choice for zener-based regulators.

188

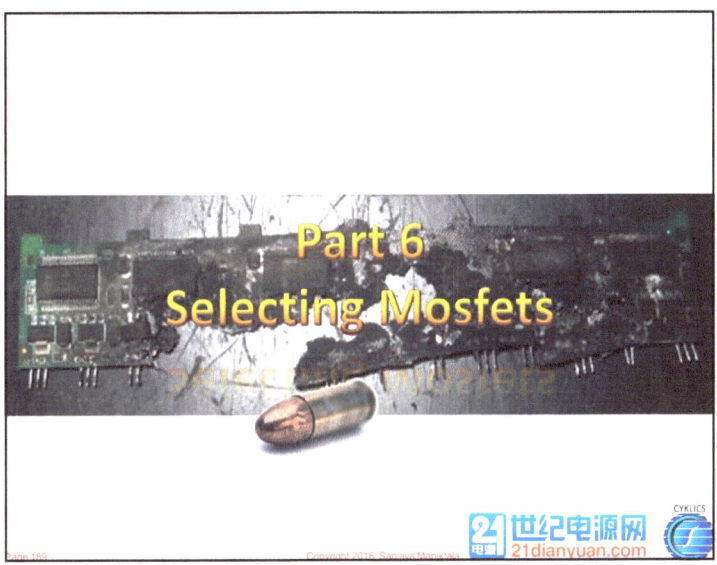

189

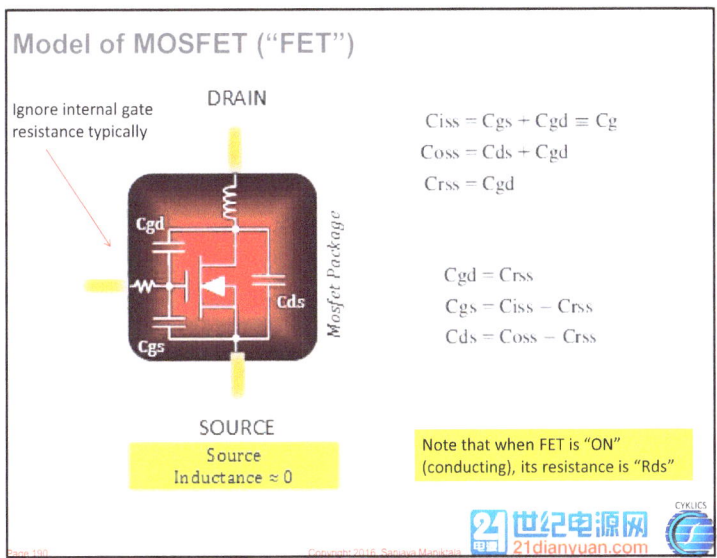

190

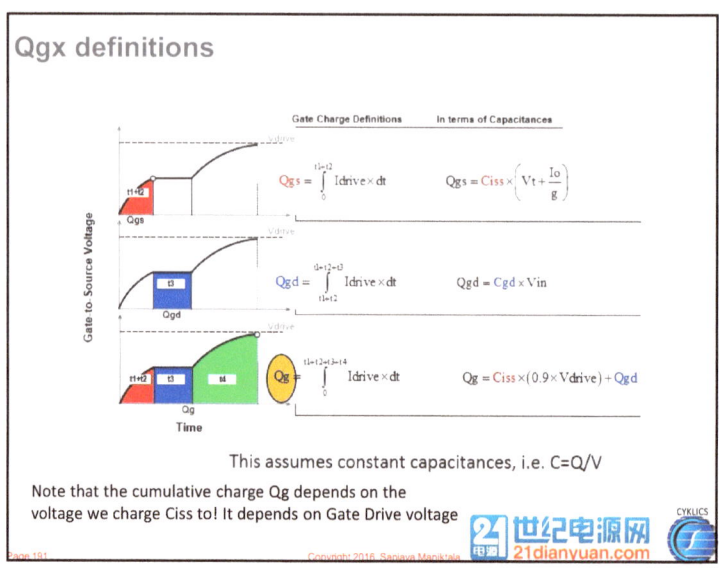

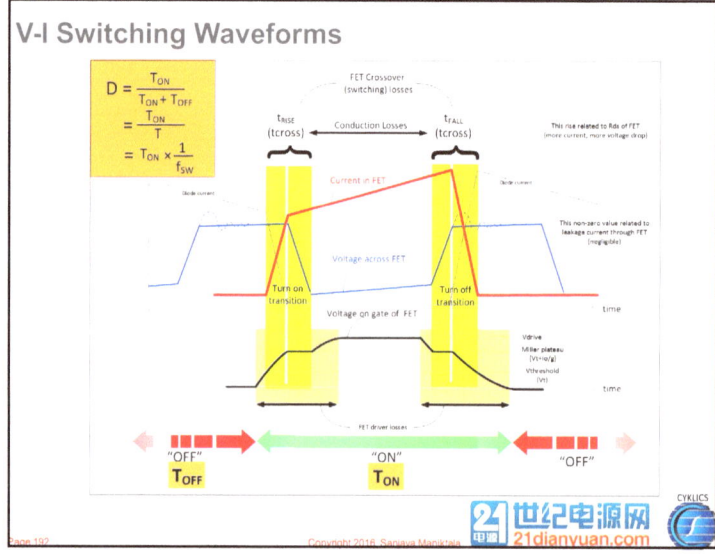

Current and voltage swings: Need for Qsw

- The crossover time is the time during which either or both the current and voltage swing.
- **During turn-ON:** The current starts swinging from a gate voltage of Vt, to the end of the Miller plateau.
- The voltage swings only during the Miller plateau.

- **During turn-OFF:** Similarly.

- So the best way to compare FETs is in terms of the charge required to be dumped in the FET for it to switch "Qsw" (not necessarily Qg as often done. Besides Qg depends on voltage of driver (overcharge region). Qgs + Qgd is a better way of comparing FETs in the absence of a specified Qsw.

193

Comparing Figures of Merit (FOM) of FETs

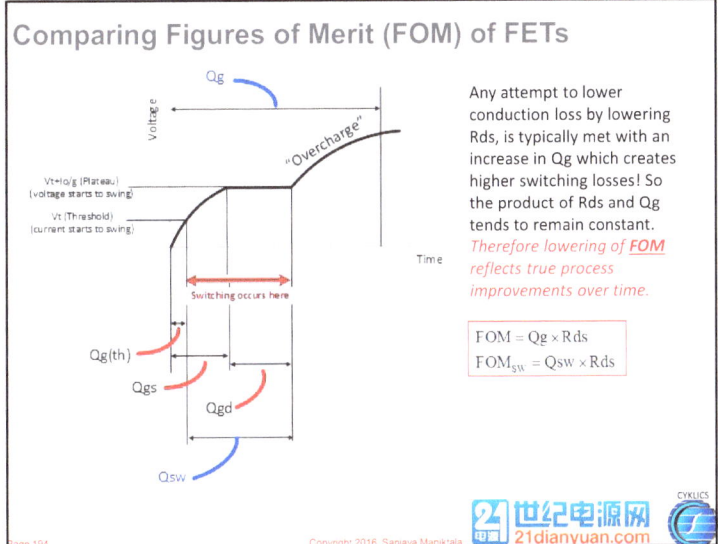

Any attempt to lower conduction loss by lowering Rds, is typically met with an increase in Qg which creates higher switching losses! So the product of Rds and Qg tends to remain constant. *Therefore lowering of **FOM** reflects true process improvements over time.*

$$FOM = Qg \times Rds$$
$$FOM_{SW} = Qsw \times Rds$$

194

Use Qg, not Ciss, to estimate Driver Dissipation

- As explained in Chapter 8 of A-Z/2e, do not use Ciss.

- We may discover that

$$\text{Ciss} \neq \frac{\text{Qg}}{\text{Vdd}}$$

- **Reason:** Typically (not always) Ciss is stated at say "15V" (for 30V FETs), but it *can be a strong function of voltage*, and at low voltages, Ciss can be much higher than stated.

In that case, we will need to either apply a scaling factor (typically x1.5) to the datasheet value of Ciss, **or better still just use Qg,** since that is the **total** gate charge --- accumulated all the way from 0V to Vdd.

Page 195   Copyright 2016, Sanjaya Maniktala

195

For example: Si4670 from Vishay

				Ch-1	680		
Input Capacitance	C_{iss}	Channel-1		Ch-2	680		
		V_{DS} = 13 V, V_{GS} = 0 V, f = 1 MHz		Ch-1	120		pF
Output Capacitance	C_{oss}			Ch-2	180		
		Channel-2		Ch-1	55		
Reverse Transfer Capacitance	C_{rss}	V_{DS} = 13 V, V_{GS} = 0 V, f = 1 MHz		Ch-2	70		
		V_{DS} = 13 V, V_{GS} = 10 V, I_D = 7 A		Ch-1	12	18	
Total Gate Charge	Q_g	V_{DS} = 13 V, V_{GS} = 10 V, I_D = 7 A		Ch-2	12	18	
				Ch-1	5.5	8.5	
		Channel-1		Ch-2	5.5	8.5	nC
Gate-Source Charge	Q_{gs}	V_{DS} = 13 V, V_{GS} = 4.5 V, I_D = 7 A		Ch-1	2		
		Channel-2		Ch-2	2		
Gate-Drain Charge	Q_{gd}	V_{DS} = 13 V, V_{GS} = 4.5 V, I_D = 7 A		Ch-1	1.5		
				Ch-2	1.5		
Gate Resistance	R_g	f = 1 MHz		Ch-1	2.5		Ω
				Ch-2	3.2		

Calculate: Qg/V = 12nC/13V = 923pF. But Ciss is stated above as 680pF…if we had used only Ciss (with no scaling factor applied), we would have underestimated the driver dissipation by a factor of 2/3 on this count alone! And the crossover time.

But also keep in mind that Qg value is only stated above for Vgs up to 4.5V. But if we are using a 10V drive instead of a 5V TTL drive, we need to increase the Qg value stated above by roughly x2 to account for the higher voltage we would charge Ciss up to every cycle!

Page 196   Copyright 2016, Sanjaya Maniktala

196

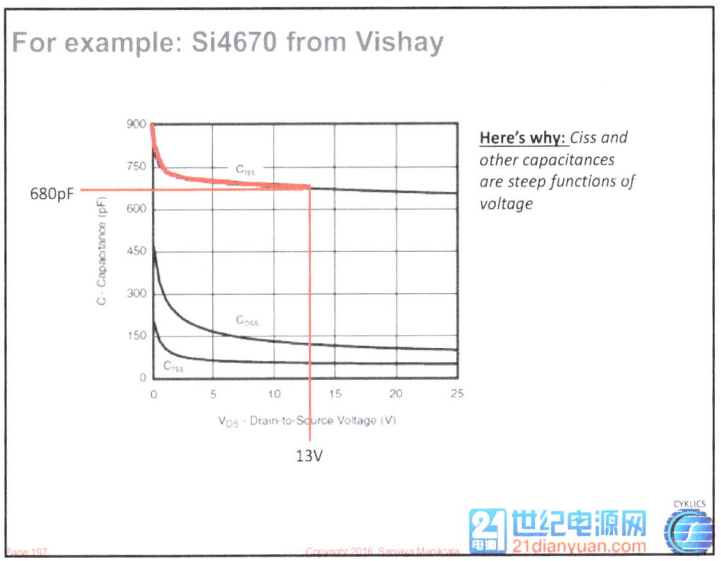

197

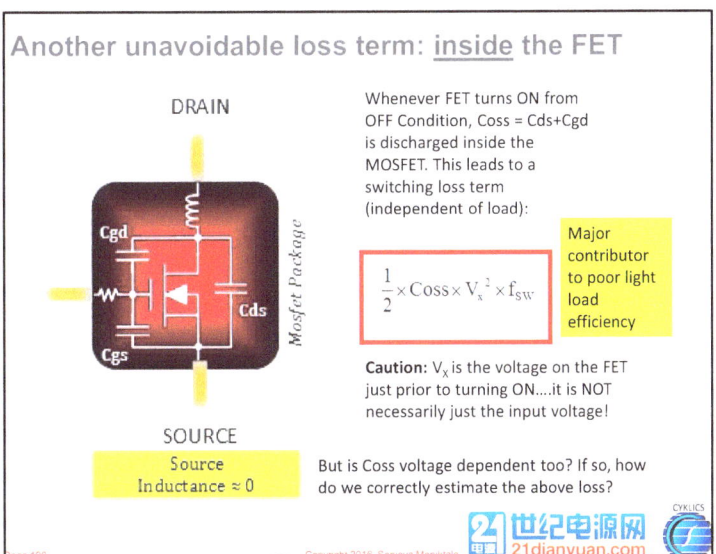

198

Infineon OptiMos 30V FET

BSC0502NSI

Input capacitance[1]	C_{iss}	-	1200	1600	pF	V_{GS}=0 V, V_{DS}=15 V, f=1 MHz
Output capacitance[1]	C_{oss}	-	420	570	pF	V_{GS}=0 V, V_{DS}=15 V, f=1 MHz
Reverse transfer capacitance	C_{rss}	-	44	-	pF	V_{GS}=0 V, V_{DS}=15 V, f=1 MHz
Gate to source charge	Q_{gs}	-	3.1	-	nC	V_{DD}=15 V, I_D=30 A, V_{GS}=0 to 4.5 V
Gate charge at threshold	$Q_{g(th)}$	-	2.0	-	nC	V_{DD}=15 V, I_D=30 A, V_{GS}=0 to 4.5 V
Gate to drain charge	Q_{gd}	-	2.3	-	nC	V_{DD}=15 V, I_D=30 A, V_{GS}=0 to 4.5 V
Switching charge	Q_{sw}	-	3.4	-	nC	V_{DD}=15 V, I_D=30 A, V_{GS}=0 to 4.5 V
Gate charge total	Q_g	-	9.0	12	nC	V_{DD}=15 V, I_D=30 A, V_{GS}=0 to 4.5 V
Output charge[2]	Q_{oss}	-	13.5	18	nC	V_{DD}=15 V, V_{GS}=0 V

Since **Coss is a steep function of voltage**, use **Qoss instead** to estimate effective (max) output capacitance:
Coss_eff=Qoss/Vdd = 18nC/15V=1200pF. Compre to Coss = 570pF above, which would have underestimated this dissipation term by a factor of ~ 1/2

199

A high-voltage CoolMos FET Datasheet

IPW65R019C7 Datasheet (Infineon)

Unfortunately, these are **all typ** values!

Input capacitance	C_{iss}	-	9900	-	pF	V_{GS}=0V, V_{DS}=400V, f=250kHz
Output capacitance	C_{oss}	-	160	-	pF	V_{GS}=0V, V_{DS}=400V, f=250kHz
Effective output capacitance, energy related[1]	$C_{o(er)}$	-	338	-	pF	V_{GS}=0V, V_{DS}=0...400V
Effective output capacitance, time related[2]	$C_{o(tr)}$	-	3320	-	pF	I_D=constant, V_{GS}=0, V_{DS}=0...400V
Gate to source charge	Q_{gs}	-	53	-	nC	V_{DD}=400V, I_D=58.3A, V_{GS}=0 to 10V
Gate to drain charge	Q_{gd}	-	71	-	nC	V_{DD}=400V, I_D=58.3A, V_{GS}=0 to 10V
Gate charge total	Q_g	-	215	-	nC	V_{DD}=400V, I_D=58.3A, V_{GS}=0 to 10V

[1] $C_{o(er)}$ is a fixed capacitance that gives the same stored energy as C_{oss} while V_{DS} is rising from 0 to 400V
[2] $C_{o(tr)}$ is a fixed capacitance that gives the same charging time as C_{oss} while V_{DS} is rising from 0 to 400V

CHECK:
a)For Driver Dissipation: Calculate Qg/V = 215nC/400V =536pF. But Ciss = 990pF above.
This time Qg will give wrong (lower) results? Instead use Ciss in P= Ciss x Vdd^2 x f

b) For FET internal Coss-based dissipation: Vendor provides **Co(er)**=3320pF. Use this, not Coss of 160pF in P= ½ $C_{o(er)}$ x V_x^2 x f

200

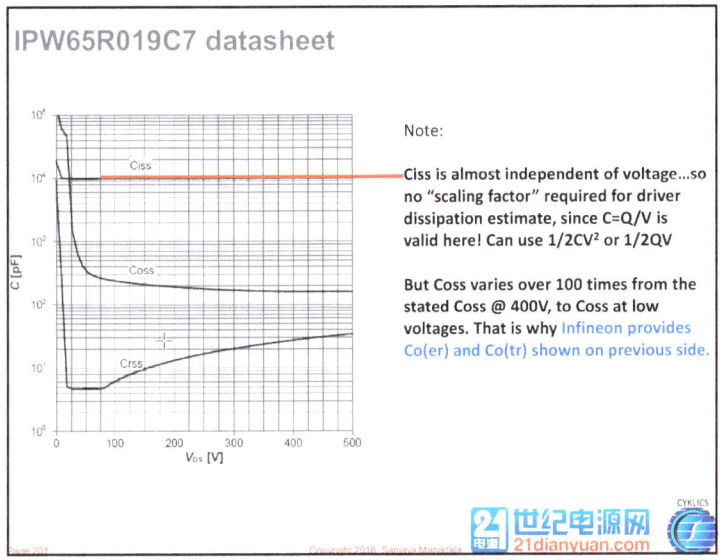

IPW65R019C7 datasheet

Note:

Ciss is almost independent of voltage...so no "scaling factor" required for driver dissipation estimate, since C=Q/V is valid here! Can use 1/2CV² or 1/2QV

But Coss varies over 100 times from the stated Coss @ 400V, to Coss at low voltages. That is why Infineon provides Co(er) and Co(tr) shown on previous side.

201

Another Valuable Lesson

• Read **datasheets** carefully. Question them. Understand test conditions, "MIN/MAX/TYP limits" etc…

Note the Ciss, Coss, Crss values provided in the Vishay datasheet were "typ" values, and therefore essentially unreliable to start with

….more on this later

202

Mosfets reaching Limits?

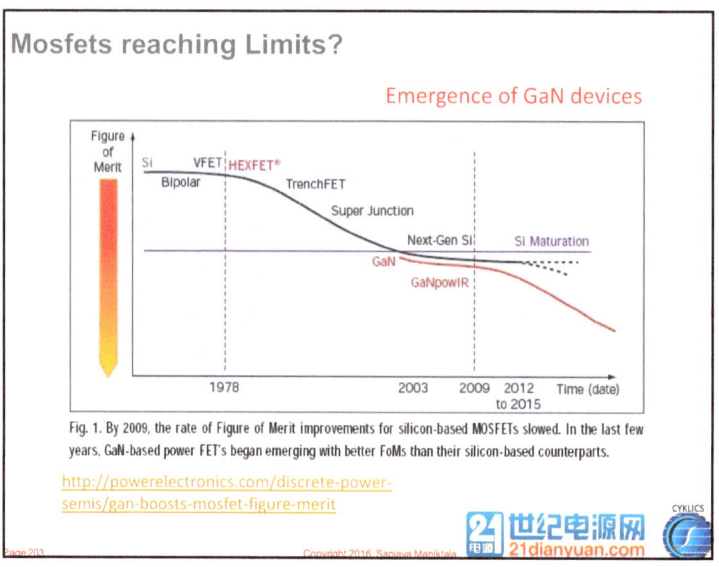

203

Emergence of GaN

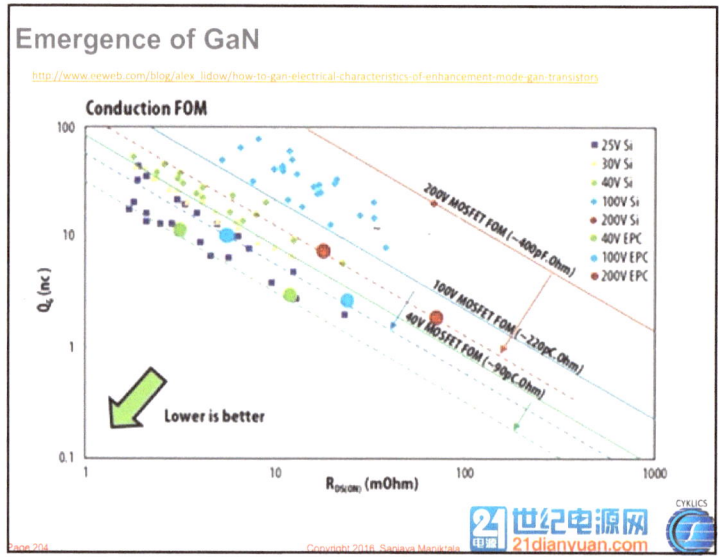

204

Comparing GaN to 30V Mosfets

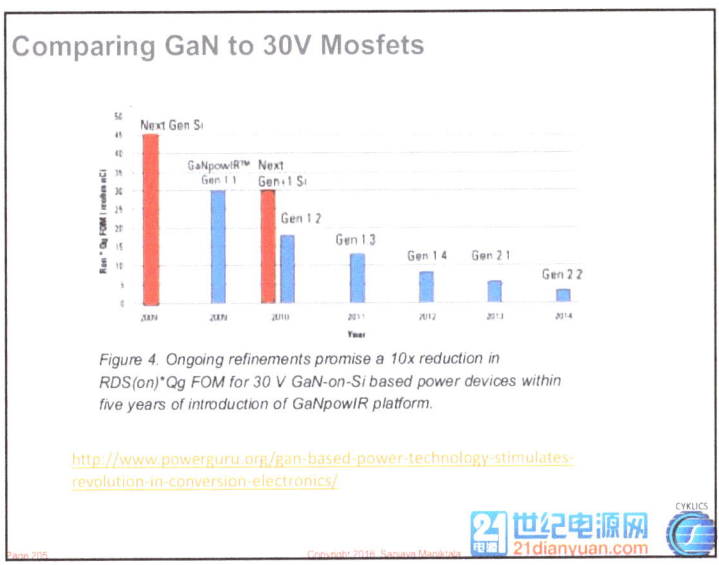

Figure 4. Ongoing refinements promise a 10x reduction in RDS(on)*Qg FOM for 30 V GaN-on-Si based power devices within five years of introduction of GaNpowIR platform.

http://www.powerguru.org/gan-based-power-technology-stimulates-revolution-in-conversion-electronics/

205

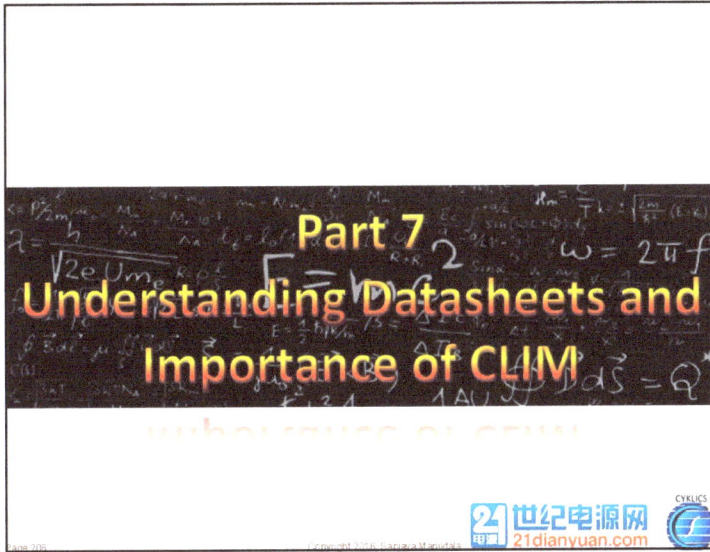

206

207

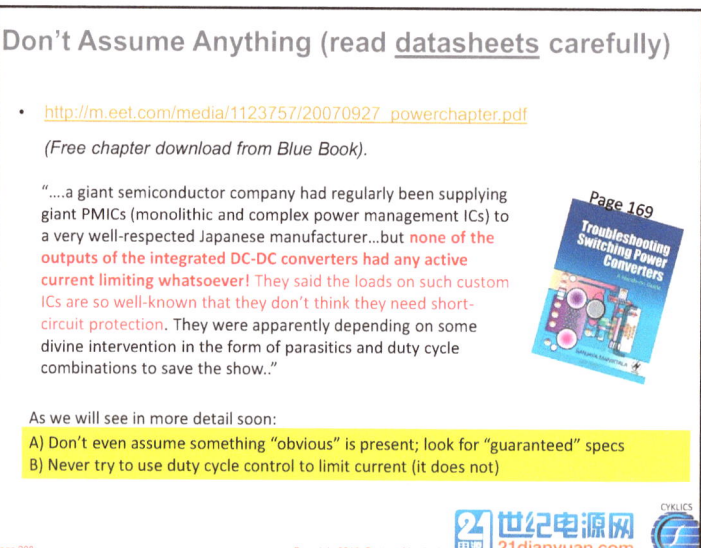

208

Understanding Datasheets

Only the Electrical Characteristics (EC) Tables constitute the legal "contract" between the vendor and the buyer.

There is no contract based on "maybe"..all "Typical" values are in reality, at best a "maybe".

25°C "MIN-MAX" limits are meaningless!

Typical Performance curves are just a "help"

Appendix F of Pease's book (pages 199 to 202) says "typical" may mean:
- The manufacturer made at least one part that met this specification.
- Half of the parts in current production are better, and half are worse.
- Once upon a time, half the parts in production were better and half were worse.
- Most parts are close to this, but some may be a little better and some may be much worse.

209

Look for Guaranteed Specs (over temperature)

Use MIN (minimum) and MAX (maximum) limits **over temperature**

- The only time I really use a "typ" value is to design the voltage (resistor) divider of a switching power supply:**"Vref (typ)"**,

  (But to know the actual variation ("spread") of **Vout,** we still need to know min and max limits of Vref over temperature).

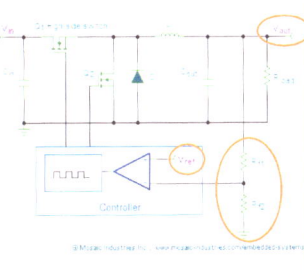

210

Critical Information Missing from Some/Most Datasheets of Switchers/Controllers

Examples where either just a "typ" value may be provided, or only "MIN/MAX at 25°C" *or as typical performance curves (all are quite useless in reality)*

- A) Current Limits
- B) Ramp Amplitude of PWM comparator
- C) Slope Compensation in Current Mode Control
- D) Switching Frequency
- E) Integrated FET Rds
- F) Internal Compensation

Ask the vendor for these, or maybe you shouldn't buy

Page 211   Copyright 2016, Sanjaya Maniktala

211

The Importance of Current Limit Spread

For "low voltage" Buck regulators, we often disregard the current limit of the device entirely. For example, we may be using a "2A switcher IC", with a set current limit of a little above 2.5A. But for a 1A application we typically select a "1A inductor" anyway, not a 3A or 4A inductor. We do this because we are simply basing the inductor selection on thermal requirements, i.e. **continuous power requirement**, not on core saturation.

But should we be looking more closely at core saturation?

Page 212   Copyright 2016, Sanjaya Maniktala

212

The Importance of Current Limit Spread

If and when we apply sudden shorts on the outputs (if accessible), the switch current will suddenly ramp up to the current limit . This can cause "core saturation", and immediate destruction of the FET.

In experiments with the LM259xHV Buck Switcher family, we realized that core saturation at "high input voltages" ("<40V") can occur so quickly that the current limit of the switcher may not be fast enough to protect the switch. The following warning was added to the datasheet/App Note AN1207...

213

The Importance of Current Limit Spread

LM2593HV Evaluation Board

National Semiconductor
Application Note 1207
Sanjaya Maniktala
July 2001

LM2593HV Evalu...

Specifications of the Board

The board is designed for a nominal DC input of 48V, but can safely withstand up to 60V. The regulated DC output is 12V at a maximum load current of 2A. It uses the Adjustable

The switch ON-time is

$$t_{ON} = \frac{D}{f} = \frac{(12 + 0.5) \times 10^6}{(48 - 1.5 + 0.5) \times 150000} \ \mu secs$$

$$L = \frac{...}{0.3 \times 2.0}^{...} \ ...$$

where

- L = 101.8 μH

(5)

The first pass selection of the inductor is usually on the basis of the inductance calculated above and the max load current. But, if the input voltage exceeds 40 V, as it does here, evaluate the inductor further to ensure that the converter withstands damage if the outputs are overloaded/shorted. A 100 μH 1.8A drum core type (large inherent air gap) was chosen from Coilcraft, which saturates above 3A. It is designed for a 40°C rise in temperature at a maximum ambient of 85°C. Its use is accepted at a load current slightly higher than its continuous rating since the maximum ambient temperature for the demo-board is only 40°C not 85°C, and since we also know it does not saturate at the maximum load current.

But even "40V" is unfortunately a typical value! For a truly reliable design, where outputs are accessible to user, we may just need to size the inductor to handle the **current limit**

214

Geometrical Center of Inductor Current Waveform

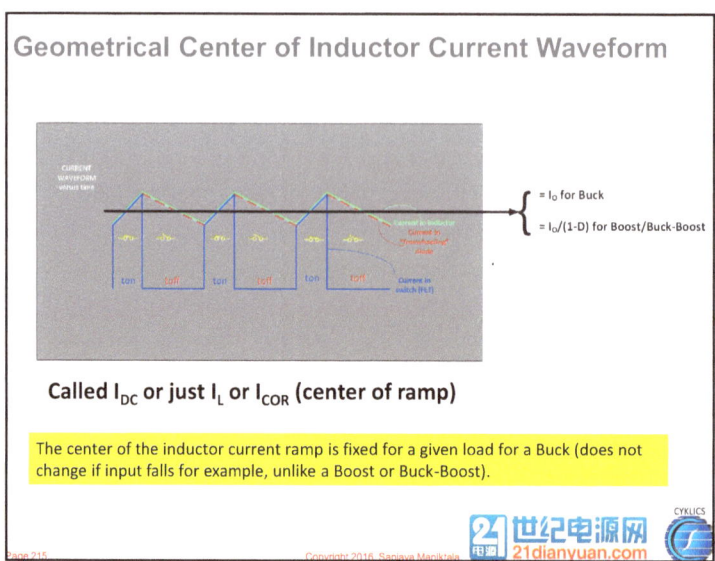

Called I_{DC} or just I_L or I_{COR} (center of ramp)

The center of the inductor current ramp is fixed for a given load for a Buck (does not change if input falls for example, unlike a Boost or Buck-Boost).

Highest Stresses, Highest Failure Rates

- Buck converters tend to fail (softly) due to high <u>load currents not at low input voltages</u>.

- Boost and Buck-Boost converters tend to fail during power-up and power down…since <u>inductor current is very high</u> (D is very high, and so 1-D tends to get close to zero, making I_{COR} very high). If input voltage is low… it can cause core saturation and immediate destruction of the FET!

- *So old computers (which typically used the Flyback topology) always seemed to fail as soon as they were plugged in….!*

Highest Stresses, Highest Failure Rates

- In Boost and Buck-Boost based topologies (such as Flyback), it is very important to set a *precise cycle-by-cycle, fast, current limit* for the switching FET, to avoid reliability issues.

-

- Also <u>careful</u> design of magnetics to avoid saturation vis-à-vis set current limits! That will kill the FET instantly.

Maintain tight tolerances on current limits! Consider worst case current limits!

217

Current Limit Spread Example

Datasheet Current Limit Spread of LM2593HV

I_CLIM	Switch current Limit	Peak Current, (Note 8) (Note 9)	3.0		A
				2.4/2.3	A(min)
				3.7/4.0	A(max)

<mark>But, should we then select the inductor as per the MIN or the MAX of the current limit spread?</mark>

218

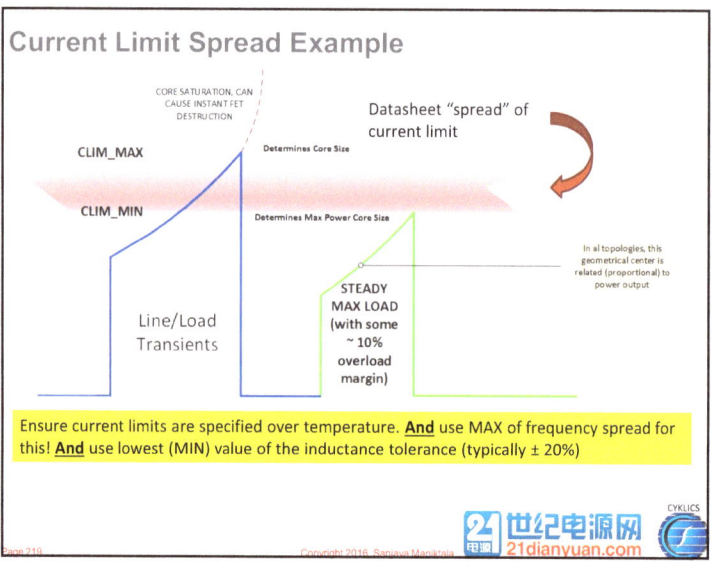

Current Limit Spread Example

CORE SATURATION, CAN CAUSE INSTANT FET DESTRUCTION

Datasheet "spread" of current limit

CLIM_MAX

Determines Core Size

CLIM_MIN

Determines Max Power Core Size

In al topologies, this geometrical center is related (proportional) to power output

Line/Load Transients

STEADY MAX LOAD (with some ~ 10% overload margin)

Ensure current limits are specified over temperature. **And** use MAX of frequency spread for this! **And** use lowest (MIN) value of the inductance tolerance (typically ± 20%)

219

Basic Power Stage Design Procedure

- 1) Know desired maximum guaranteed power ("max load").
- 2) Add 10% to add that as "overload margin" to handle normal line (input)/load transients at max load.
- 3) Select inductance based on max load and desirable current ripple ratio "r" of 0.4 (*Maniktala*)...
- 4) Draw out the resultant switch/inductor current waveform. 5) Determine peak value of that waveform, and set MIN of current limit at this peak value (if adjustable).
- 6) Select core size/inductor "Isat" rating based on the MAX of current limit.
- 7) However, wire gauge is selected based on thermal considerations… i.e. based on max load (~MIN of current limit) since that is the continuous maximum power

220

Selecting Integrated Controllers carefully-1

- These integrate the FET and control for convenience. They usually set a **fixed** current limit internally, based on a) the perceived continuous current capability of the FET…and also b) vague thermal assumptions (which are hard to replicate anyway). They then create a "family" of parts.

- **Advice:** Do not blindly trust their "maximum power" front page numbers…this is likely just their marketing department at work. Calculate, calculate, calculate… (due diligence)!

- **Important:** you may not be able to set a current limit based on YOUR maximum power requirement. You need to "fit in".

221

Tight Current Limits matter. BUT…

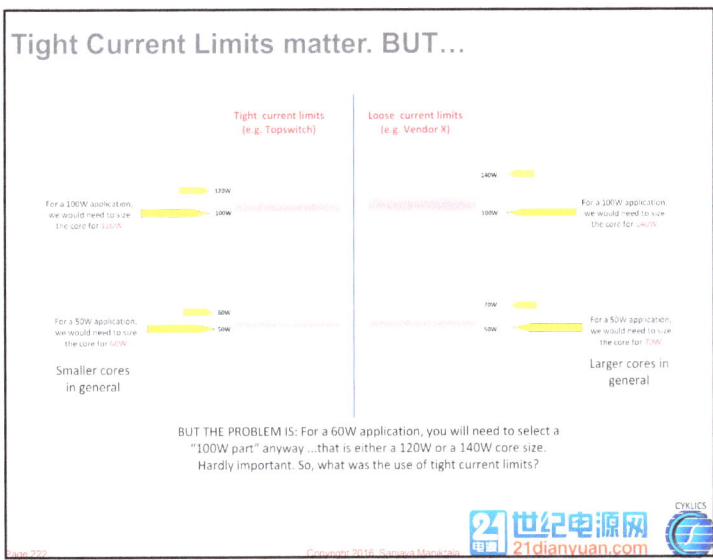

Tight current limits (e.g. Topswitch) Loose current limits (e.g. Vendor X)

For a 100W application, we would need to size the core for 100W

For a 50W application, we would need to size the core for 60W

Smaller cores in general

For a 100W application, we would need to size the core for 140W

For a 50W application, we would need to size the core for 70W

Larger cores in general

BUT THE PROBLEM IS: For a 60W application, you will need to select a "100W part" anyway …that is either a 120W or a 140W core size. Hardly important. So, what was the use of tight current limits?

222

Topswitch (Power Integrations) example

- They released a family of integrated Flyback ICs. They did the right thing: they understood the importance of tight current limits. Otherwise you could end up oversizing your core for the same power.

- But by offering a set of fixed current limits, the purpose of tight current limits was defeated, unless your application matched their part, not the other way around!

- I wrote about this in Design and Optimization and A-Z/1e.

Page 223   Copyright 2016. Sanjaya Maniktala

223

Page 88 of A to Z First Edition

On the other hand, manufacturers of *off-line* switcher ICs *do* need to maintain a *tight tolerance* on the current limit. In their case, the maximum power-handling capability of their particular device is in effect dependent only on the 'MIN' (minimum limit) of the current limit specification, whereas, the transformer size is determined entirely by the 'MAX' of the current limit specification. So in this case, a "loose" current limit specification effectively amounts to requiring *bigger components* (transformer) for the same maximum power-handling capability.

Note: Some makers of off-line integrated switcher ICs (e.g. the "Topswitch" from Power Integrations) often tout their "precise" current limit — thus suggesting that we get the best power-to-size ratio (i.e. converter power density) when using their products. However, we should remember that in most cases, their product families have a *discrete* set of fixed current limits. And that is a problem! For example, we may have devices available with current limits in steps of 2 A, 3 A, 4 A, and so on. So yes, we may indeed get a higher power density *when operating at the maximum rated output power* of a particular IC. But when operating at a power level *between* available current limits, we are not going to get an optimum solution. For example, in an application where the peak current is 2.2 A, then we would need to select the 3 A current limit part, and we will need to design our magnetics to avoid core saturation at 3 A. So in effect, we have a very imprecise current limit now! The best solution is to look for a part (integrated switcher or controller plus mosfet solution) where we can precisely set the current limit *externally*, depending upon our application.

Page 224   Copyright 2016. Sanjaya Maniktala

224

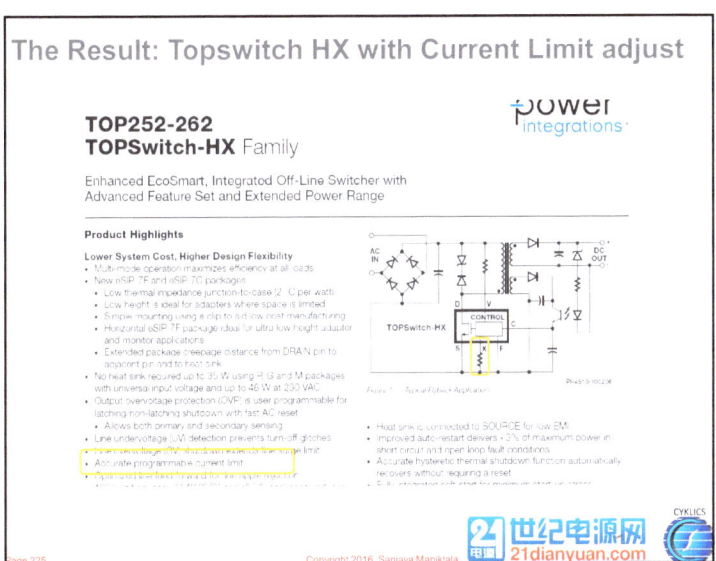

225

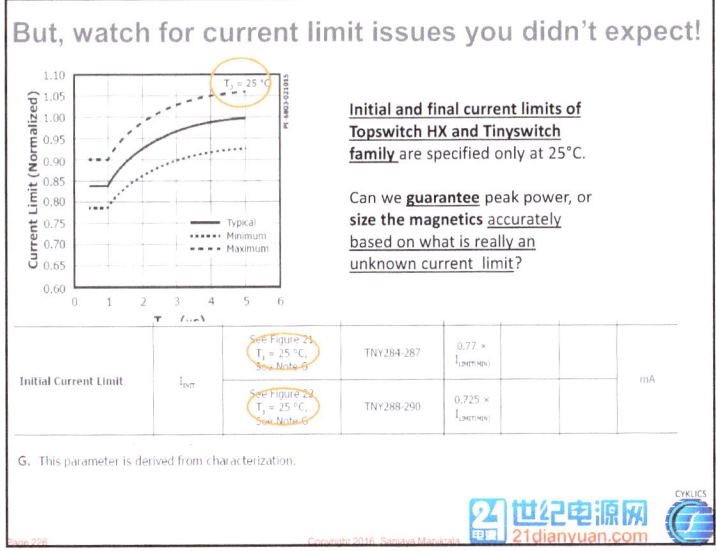

226

Other current limit issues you didn't expect!

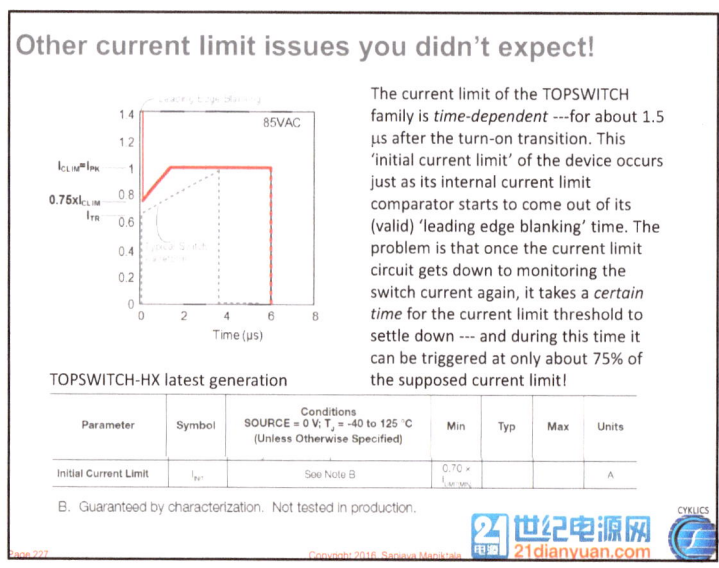

The current limit of the TOPSWITCH family is *time-dependent* ---for about 1.5 µs after the turn-on transition. This 'initial current limit' of the device occurs just as its internal current limit comparator starts to come out of its (valid) 'leading edge blanking' time. The problem is that once the current limit circuit gets down to monitoring the switch current again, it takes a *certain time* for the current limit threshold to settle down --- and during this time it can be triggered at only about 75% of the supposed current limit!

TOPSWITCH-HX latest generation

Parameter	Symbol	Conditions SOURCE = 0 V; T_J = -40 to 125 °C (Unless Otherwise Specified)	Min	Typ	Max	Units
Initial Current Limit	I_INIT	See Note B	0.70 × I_LIM(MIN)			A

B. Guaranteed by characterization. Not tested in production.

227

228

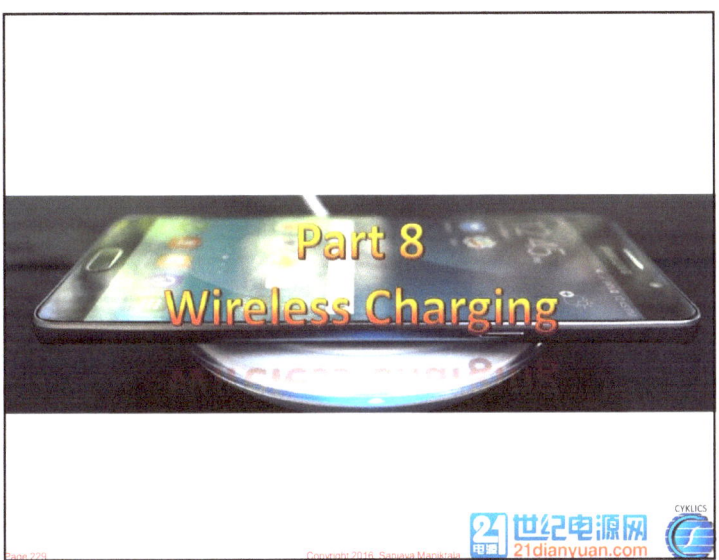

229

The Game

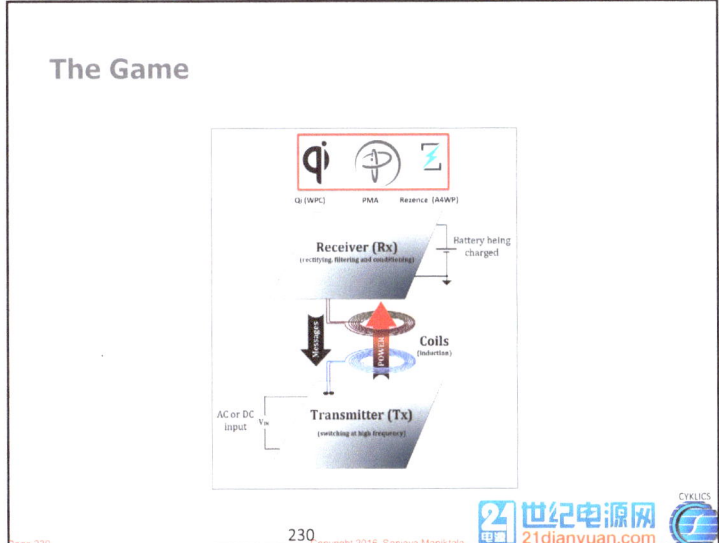

230

The Teams

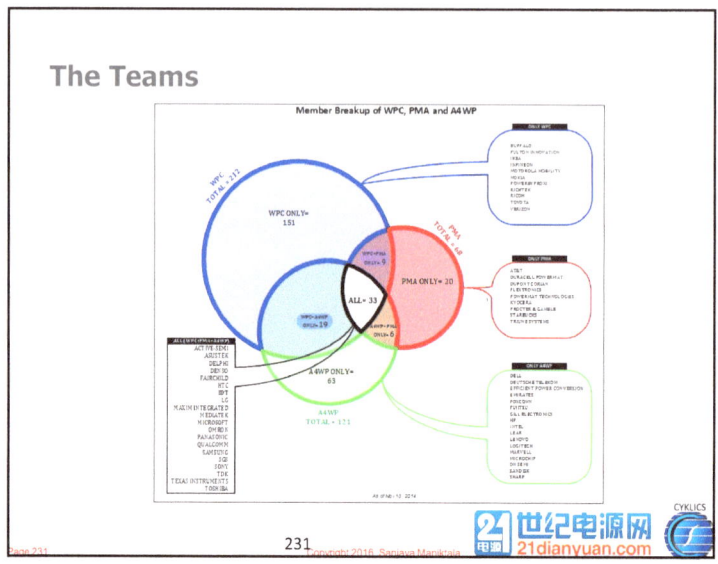

231

Switch a Voltage Across a Coil

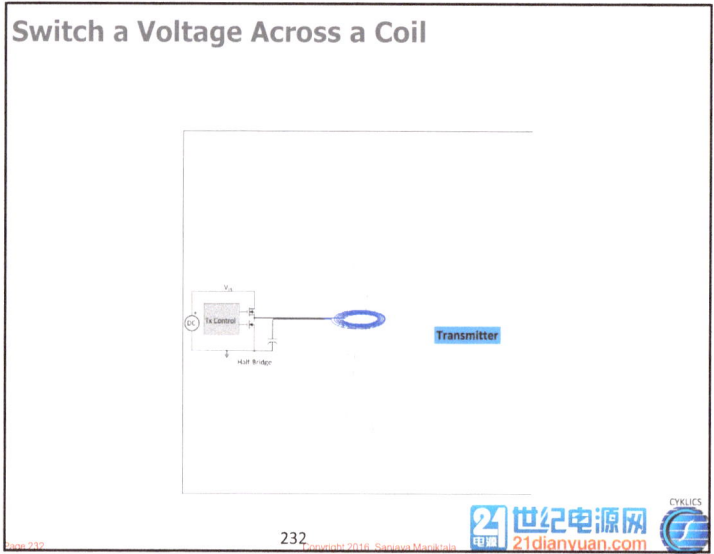

232

Add Another Coil to Pick up the Transferred Energy

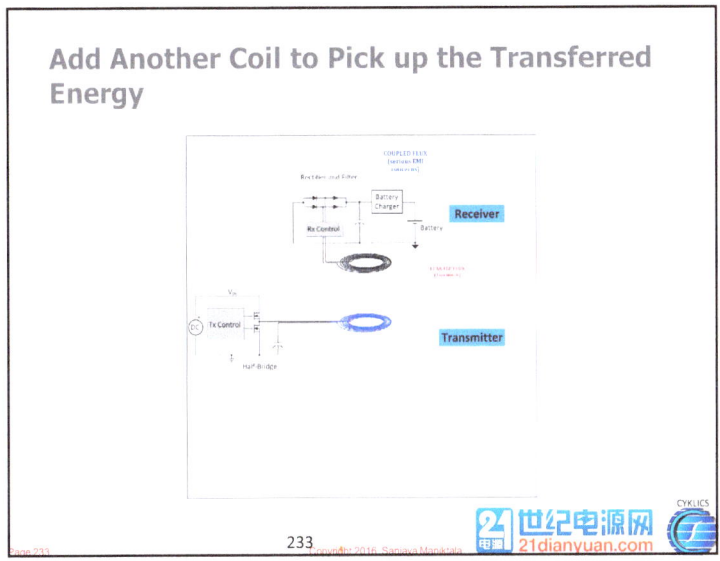

233

Add Ferrites to Enhance Coupling

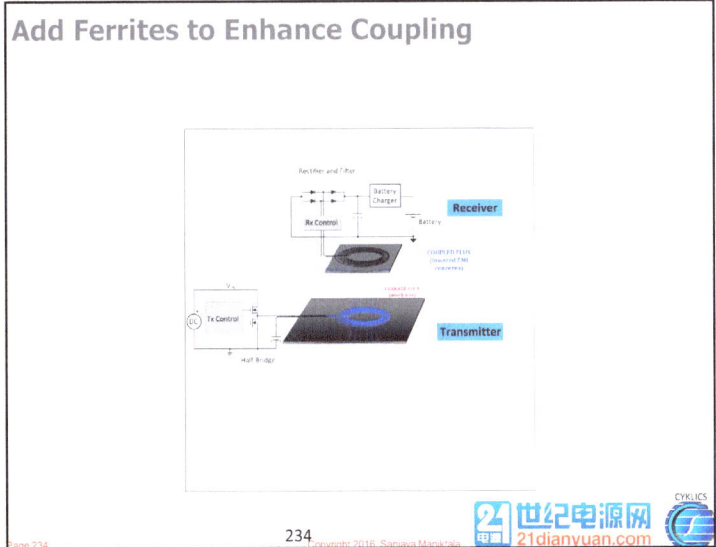

234

Close the Feedback Loop (Communicate using backscatter)

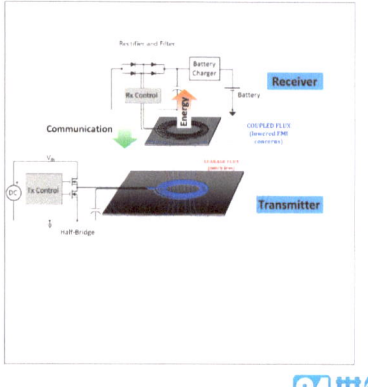

235

Highlights

Leakage is still present, but being used to our advantage.

Leakage energy is <u>recovered</u> by use of resonance.

We need to operate the resonant circuit in "inductive" mode (i.e. to the right of the resonance peak) and allow more deadtime (>200ns) for the switching node to rise to the rails by itself, before the FET is turned ON. This is called Zero Voltage Switching (ZVS). Then, we have almost no switching losses (only higher conduction losses (but we can use low Rds FETs to reduce that).

SAME BASIC PRINCIPLE AS THE LLC TOPOLOGY. Only the transformer looks different!

236

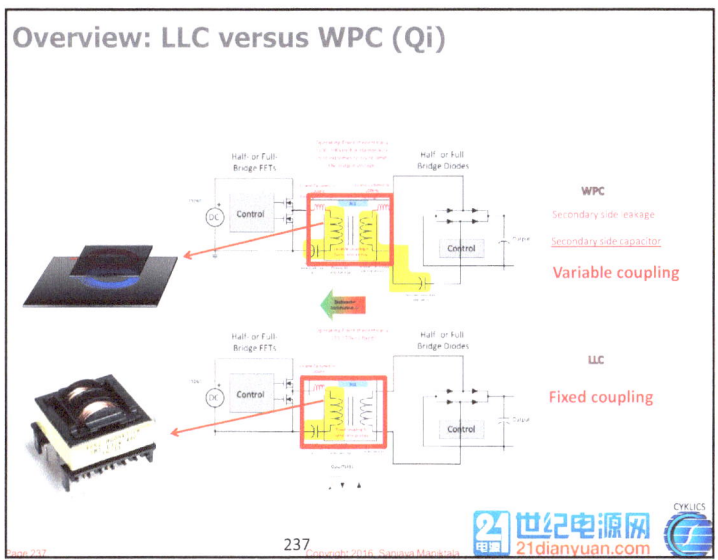

237

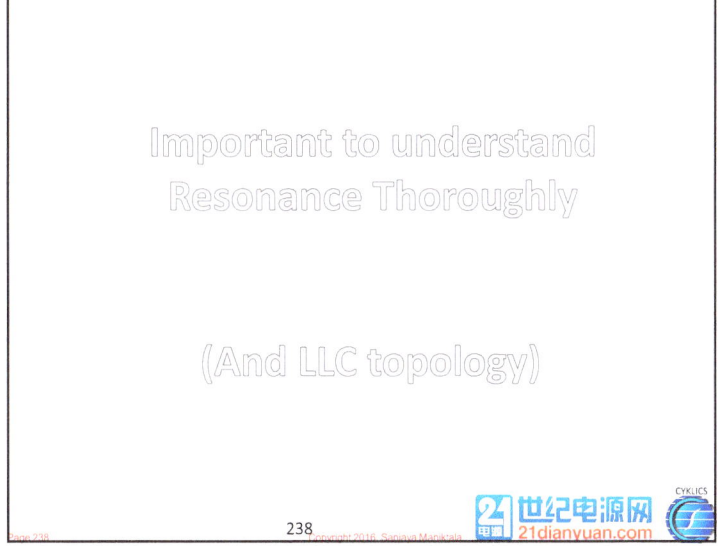

238

Parallel Resonance Explained

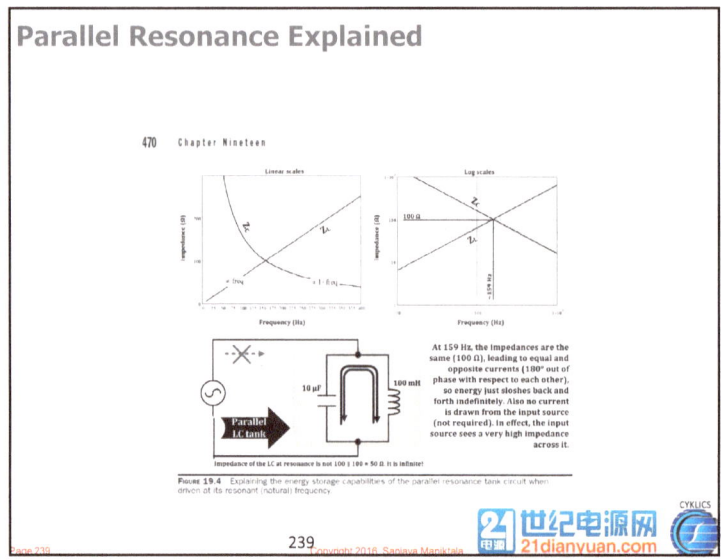

239

LLC+ Resonance: References

240

Desirable Attitudes to Succeed as a Power Designer

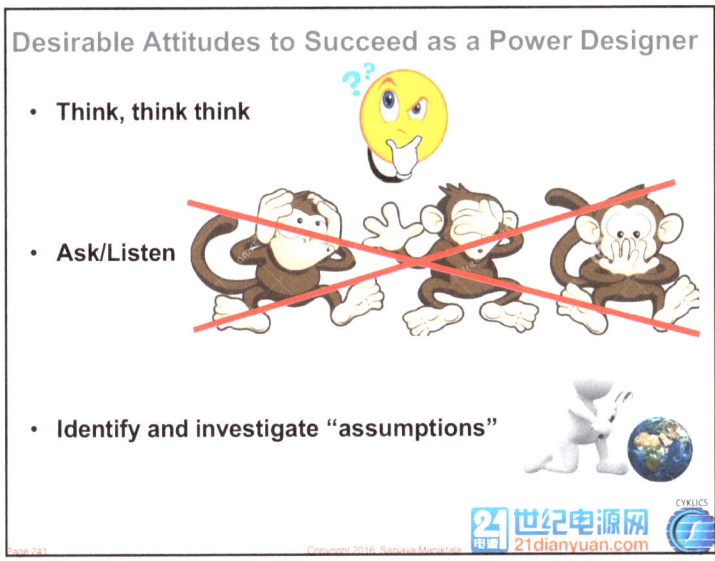

- **Think, think think**

- **Ask/Listen**

- **Identify and investigate "assumptions"**

241

The solution to avoid huge future problems

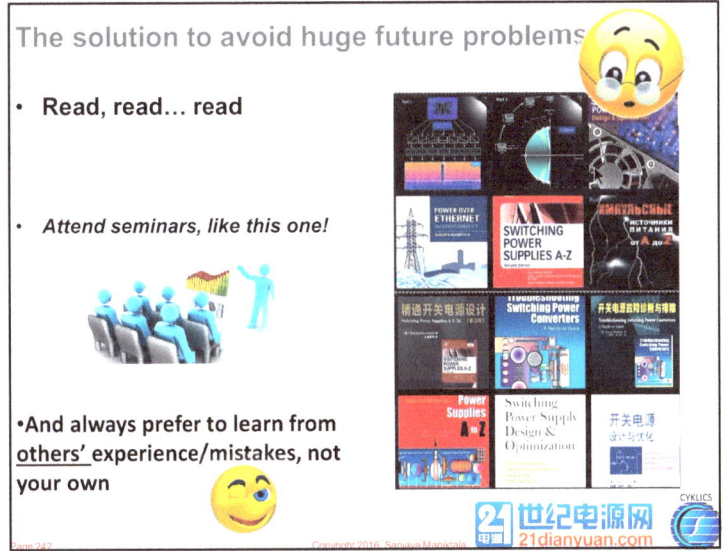

- **Read, read… read**

- *Attend seminars, like this one!*

- And always prefer to learn from <u>others'</u> experience/mistakes, not your own

242

Acknowledgments

- This seminar wouldn't have been possible without the incredible help and support from (in order of contact):

- A) Eric Wen
- B) Mr. Lei Liang
- C) Ms. Carolyn Lin
- D) Lin (Tony) Dong

- Also, I need to thank Eric Wen once again, for valuable help rendered in structuring this seminar just right for you today, and for his great help in translations throughout.

243

Thank you my dear friends from China! God bless you! Good luck designing much better power supplies soon!

Go to www.cyklics.com and post comments or contact me in future

244

www.ingramcontent.com/pod-product-compliance
Lightning Source LLC
Chambersburg PA
CBHW040904180526
45159CB00010BA/2920